The Pioneer Period of Hungarian Airmail

by

Victor G. Berecz, Jr.

SPECIAL NOTICE

Permission is hereby given for use of background data contained in this book, and covered by copyright, by publishers in newspapers, magazines, and periodicals and by airpost dealers in advertising matters, circulars or price lists prepared and issued for free distribution, provided proper acknowledgment of the source of the material so used is given in each newspaper, magazine, periodical, advertising matter, circular or price list. All other unauthorized use, including verbatim reproduction, is strictly forbidden except by specific permission.

Library of Congress Catalog Card Number: 96–85087

ISBN 0–939429–16–0

Cover Photos

Front: an *Aeroexpress RT* seaplane at its dock on the Danube, near the Elisabeth bridge, in 1924.

Back: (a) the airship of Dávid Schwarz preparing for takeoff at Templehof (Berlin) in 1897; and (b) men and equipment of *Fliegerkompagnie 11* at Przemysl during the winter of 1914–15.

Table of Contents

American Air Mail Society

Organized 1923 as the Aero Philatelic Society of America

About the Author

Vic Berecz has specialized in Hungarian aerophilately for over thirty years. He is a member of the APS, AAMS, AAPE, the Hungarian Philatelic Society of Great Britain, and is a founding member and former President of the Society for Hungarian Philately.

Mr. Berecz began exhibiting Hungarian airmail in 1966. His exhibits typically cover the pioneer period (through 1918), international airmail to and from Hungary during the 1922–27 period, or Zeppelin mail of Hungary. Though he usually exhibits only once each year, between 1974 and 1994 he received 19 gold awards at WSP exhibitions in the U.S. His collection has been shown internationally on only a few occasions, typically garnering a vermeil award. Mr. Berecz' aerophilatelic publications include a 1970 monograph entitled *The Pioneer Period of Hungarian Airmail*, and about 40 articles published over the years in *News of Hungarian Philately*, *Stamps of Hungary*, *Hungarian Heritage Review*, the *Airpost Journal*, *Linn's Weekly Stamp News*, and the *American Philatelist*.

Professionally, Mr. Berecz is a software engineering manager. He retired from a long career at United Technologies Corporation in 1994, where he last served as Manager of the Advanced Computing Technology Center. He also served as chairman of the Technical Advisory Board, and member of the Board of Directors, of the Software Productivity Consortium. In his retirement, Mr. Berecz is an adjunct lecturer at the Hartford Graduate Center, and pursues his hobbies of philately, genealogy, and gardening.

PREFACE

The objective of this book is to provide an enduring and definitive reference work on the first fifty years of Hungary's aeronautical and aerophilatelic history. To provide the desired depth of detail, I consulted a comprehensive set of source material. An annotated bibliography of those sources is provided at the end of the book. Augmenting the standard literature was my 30+ years of collecting and personal research, specializing in Hungarian airmail. Finally, I *let the material speak for itself.* This book is illustrated with over 300 figures, and the great majority of them are Hungarian airmail covers from my collection ... which has repeatedly won gold awards at WSP exhibits.

The text of the book is divided into two parts. The short, initial part (Section 1) provides the historical context. It is essentially a brief overview of the political and social history of Hungary, focusing on the five decades from 1896 to 1946. Issues that bear upon the development of aviation and airmail in Hungary are emphasized. This part of the book is primarily intended for readers unfamiliar with Hungarian history.

The second part (Sections 2–6) provides thirty–one short, largely self–contained chapters on specific topics of aerophilatelic interest. It is illustrated with maps, historical photographs, and airmail material from my collection. I have grouped these chapters roughly into decades, and arranged them essentially in chronological order. As a whole, they cover the entire five decades. The focus, though, is on the earlier years, the period through 1927. Since 1966, in my exhibiting and writing, I have applied the name *The Pioneer Period of Hungarian Airmail* to those decades.

One goal of this book is to provide information on early Hungarian airmails to the English–speaking aerophilatelic community. There are many writings on the subject in Hungarian and German. I'd like to think that none of them provides, in one place, such a comprehensive picture of Hungarian aerophilately. Some philatelists may feel that being written in English limits the usefulness of this book. Certainly Hungarian and German translations would be worthwhile, and I would encourage anyone interested in such a project to contact me or the AAMS. The availability of the book's text in computer–readable form makes such a project easier to accomplish as automated translation aides improve.

I would like to dedicate this book, with thanks and many fond memories, to the three philatelists who most influenced my collecting, my research, and my exhibits of *The Pioneer Period of Hungarian Airmail* ...

Earl F. Blinn, **John J. Latham**, and **Otto Gábor Schaffling**.

Historical Context of Hungarian Aerophilately

This first part of my book attempts to provide a historical context for Hungarian aerophilately. Readers unfamiliar with Hungarian history will be the primary beneficiaries. Obviously, this overview concentrates on the fifty years from 1896 to 1946 that is the period of the book's aerophilatelic focus. This section provides a cursory review of earlier and more recent events. For those interested in further English–language reading on the general history of Hungary, I would recommend Denis Sinor's 1959 **History of Hungary**. His focus is the important people and issues; and, I believe, he takes an objective historical perspective without political bias.

A Brief History of Hungary to 1867

The Hungarians (or *Magyars*, as they call themselves) were a nomadic people who, with their horses, migrated for centuries on the plains of southeastern Europe north of the Black Sea. In 896 AD they entered the underpopulated and largely unprotected Carpathian basin of central Europe. Initially, Magyar raiding parties induced fear in the inhabitants of northern Italy, Moravia, and Bavaria. Yet, within a century, and for the first time in their history, the Magyars settled into an agricultural society that later became Europe's bulwark against other invaders from the East.

The Hungarian language was unrelated to the known European languages of the time. Hungarians wrote in a runic script. The Magyar tribes practiced an animistic religion, and in their migrations had come into close contact with many of their neighbors. As a result, for the era they were surprisingly tolerant of foreigners in their midst, and of the religious practices of others. This tolerance, as well as other factors, led to a nation of considerable religious and ethnic diversity. That diversity, coupled with the waning and waxing of the nation's tolerance over the centuries, created many problems that reached critical mass only in 1918.

Under the leadership of Géza, the Magyars moved clearly into the western orbit in the last quarter of the 10th century. This included the adoption of Roman Catholicism and close associations with the Holy Roman (i.e. German) Empire. This westernization culminated in the reign of Géza's son ... known to us as Hungary's first king and saint ... Stephen (in Hungarian István). He was invested, with a crown sent by Pope Sylvester II, on Christmas Day of the year 1000 AD. During Stephen's reign, he consolidated power, stifled reactionaries, and continued the conversion of Hungary to Christianity. Though his only son died at a young age, his successors from the Árpád dynasty ruled Hungary for 300 more years. During this period, Hungary often had to create a balance between the eastern and western empires and the Roman church. So, the nation's fortunes had "ups and downs," but Hungary was widely recognized as one of the great powers of Europe.

By 1091, King László had conquered Croatia and brought Hungary to roughly the borders it retained for over 800 years (except during the Turkish occupation). The pre–WWI map of Europe in Figure 1–1 shows those borders. [Actually, this map is quasi–philatelic. It is taken from the back of a 1909 *Grand Hotel Hungaria* envelope mailed from Hungary to the U.S.]

In addition to modern Hungary, historic Hungary included the following areas: a small part of southern Poland; all of Slovakia; the eastern portion of Austria; a small part of northeastern Slovenia; most of Croatia; northern Serbia; northwestern and central Romania; and part of western Ukraine.

In 1241, the Mongols invaded Europe and Hungary, under Béla IV, took the brunt of the attack. No European power could resist the Mongol armies in the field, and Hungary was decimated. Though the Mongols withdrew within two years, much of Hungary north and east of the Danube was both destroyed and depopulated. Hungary made a truly remarkable recovery under Béla, but during this era the seeds of future problems were sown. It was during this reconstruction that the great barons again increased their power, and the nation welcomed large numbers of Slavic, Romanian, and other settlers. Hungary's multi–ethnic character owes much of its being to the Mongol invasions.

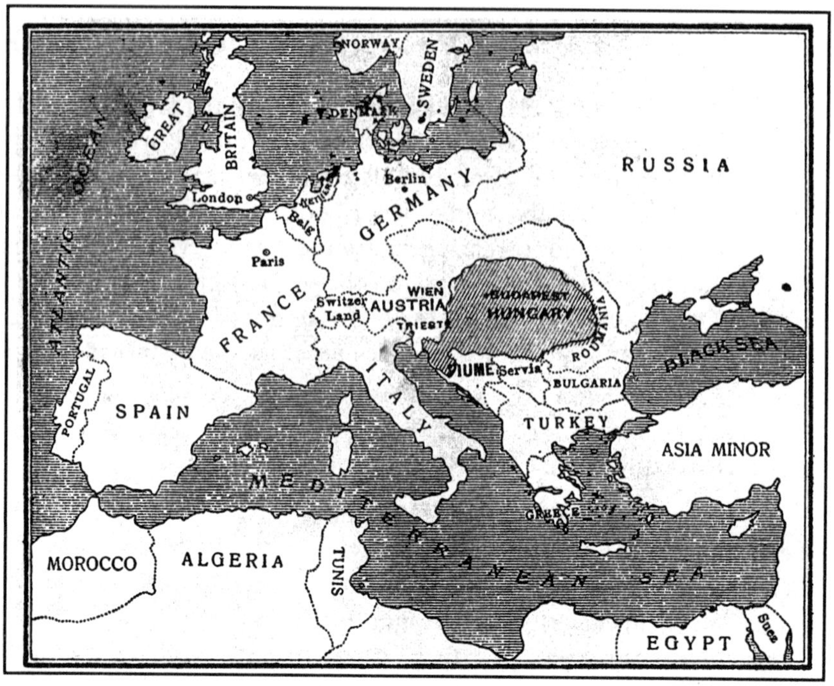

Figure 1–1 Map of Historic Hungary

The zenith of Hungarian power came during the reign of the elected monarchs who followed the extinction of the Árpád dynasty in 1301. Also during this time, a small number of wealthy barons gained control of over half the land in Hungary. In 1437, Albert was the first Habsburg elected king of Hungary, but no precedent was set ... yet. In 1453 a great catastrophe befell Europe. Constantinople, which for a millennium had been the chief city of the Eastern Roman Empire fell to the Ottoman Turks. The high point of Hungarian medieval culture came during the reign of King Matthias who not only repulsed the Turks, but brought both humanism and justice to Hungary.

The nation's well–being did not last long after Matthias' death in 1490. The Turks continued their conquest of the Balkans and finally annihilated the Hungarian army at Mohács in 1526. King Louis II succumbed in that battle. Following Mohács, Hungary was divided into three parts. The Ottoman Turks occupied the largest part of the nation. Habsburg rulers, who now took on a *de facto* hereditary role, ruled the north and west of Hungary. Transylvania functioned nominally as an independent principality, though often, of necessity, it professed allegiance to the Turks.

In western Europe, this was known as the "reformation era." The Protestant Reformation also had a major effect on the religious life of divided Hungary. The Evangelical (Lutheran), Reformed (Calvinist) and Unitarian churches all gained significant numbers of adherents. These new Protestants in Transylvania and the occupied areas did not face the threat of the counter–reformation. The three Protestant groups –– coupled with the majority Roman Catholics, the Eastern Orthodox adherents among the Slavic and Romanian minorities, and the large number of Jews who immigrated to Hungary later –– added religious diversity to the already ethnically diverse Hungarian nation.

With the failure of the 1683 Turkish attack on Vienna, the pendulum swung quickly to the side of the Christian nations. In 1686, a truly international army recaptured Buda ... which after 145 years of Turkish occupation was a smoldering ruin. The long–term results of the Turkish occupation were the depopulation of major areas and allowing good farmland to go to ruin as either deserts or marshes. In the late 1400's, Hungary had 4 million of Europe's 80 million total population. At the end of the 1600's it had but 2½ million people of Europe's 130 million. Before the occupation, Magyars were over 80% of Hungary's population. Afterwards, at 45%, Magyars were a minority in their own country.

Already before the reconquest, many Serbs settled in the deserted lands of southern Hungary. After the conquest, many Germans settled in the cities throughout Hungary. The large influx of Romanians to Transylvania continued unabated throughout the 1700's. All these added to the ethnic mix. The reconquered lands were sold, further building the estates of the great landowners since they were the only ones who could afford the price.

Ferenc Rákóczi, in the first decade of the 18th century, led Hungary's first major revolt against the Habsburgs. Upon the accession of Joseph I to the Habsburg throne, Rákóczi demanded Hungary's constitutional right to elect its monarch. After the defeat of Rákóczi, Hungary stabilized under Habsburg rule and the nation slowly began to rebuild from the Turkish occupation. Actually, Hungarian support was an important factor in the accession of Maria Theresa to the Habsburg throne when the opposition of numerous pretenders led to the War of the Austrian Succession.

The first part of the 19th century was a period of great national revival in Hungary of agriculture, industry, literature, and the arts. This took place in a political atmosphere of Habsburg opposition to any progress coupled with rekindled Magyar national aspirations. Pragmatism required the Habsburgs to reconvene the Hungarian Diet for the first time in 1825. An era of great patriotic feeling emerged in Hungary among all peoples from nobility to peasants. This set the stage for 1848, a year of turmoil across all of Europe.

Revolution in France inspired unrest in Austria. While the Habsburgs were attempting to fix their own problems in Vienna, Louis Kossuth sought to transform Hungary into a modern state. On April 11, Ferdinand V consented to the demands of the Hungarian Diet. Even Kossuth, at this point, didn't propose to sever all links with Austria. A struggle throughout the Habsburg lands ensued between moderate and extremist political leaders. Austrian extremists instigated a rift between the Magyars and their Croat compatriots. This led to civil war within Hungary by September.

In October 1848, with a full–scale revolution under way in Vienna, the Hungarian Diet refused to enforce an imperial decree and Kossuth became the virtual ruler of Hungary. Austrian reaction was quick. Ferdinand was dethroned; his younger brother was passed over; and the political leadership of Austria installed the young, malleable Francis Joseph as Emperor. Austria then began its effort to crush the Hungarian revolt. The government fled to Debrecen, and the Austrians captured Buda on December 5.

The Austrian commander prematurely declared the end of the Hungarian rebellion, which led to an imperial decree incorporating Hungary into Austria as a mere province. When the military situation stabilized somewhat, extremists in Kossuth's entourage convinced him to declare the Habsburg's dethroned.

This act was largely symbolic, for Hungary's fate was sealed. Francis Joseph had appealed to Tsar Nicolas I to help him put down the rebellion. The Austrian and Russian armies, totaling 370,000 soldiers and 1,200 guns, trapped between them the 150,000 Hungarians armed with only 450 artillery pieces. It was no match. Kossuth fled to Turkey leaving Görgei, Hungary's best general, in power. He had three choices: to flee also; engage in a last, hopeless battle; or lay down his arms. He chose to surrender to the Russians on August 13, 1849.

At that point, seventeen years of Austrian absolutism began for Hungary. During those years, Austrian international reverses left the Empire greatly weakened. Finally, the disastrous loss of the Austro–Prussian war in 1866 required a compromise with the Empire's "minorities." The largest and most powerful of those minorities were the Magyars. Austria needed the active good will and support of the Empire's Hungarians to survive.

Hungary in the Dual Monarchy

The period of absolutism was not all bad for Hungary. The integration of Hungary and Austria brought much Austrian capital to Hungary due to the lower wages there. This led to a major surge of industrialization, particularly in the food processing industries. The Hungarian rail network grew rapidly, and Budapest became a great river transport center. Also, Austrians are not known as an oppressive people, and as Sinor notes, they are "too inefficient to be persistently ruthless." On the other hand, despite Hungary's love of Kossuth who was still fomenting rebellion from exile, most Hungarians wanted to remain within the Empire and make peace with the Emperor. So, a "compromise" -- in German an *Ausgleich* -- was crafted.

The *Compromise of 1867* established the dual monarchy, commonly called the Austro-Hungarian Empire ... a unique experiment in government that lasted fifty years. The terms of the compromise were negotiated on the Hungarian side by Ferenc Deák and Gyula Andrássy and on the Austrian side personally by Emperor Francis Joseph. The Empress Elisabeth, who sympathized with the Hungarian cause, did much to smooth things over.

Figure 1-2 shows a map of Hungary as it existed within the dual monarchy during the years 1867-1918. Note that the Austrian territories to the west of Hungary included South Tyrol, Slovenia, and Istria. To the north of Hungary they included Bohemia and Galicia. The Austrian territories to the south of Hungary were Austrian Dalmatia, Bosnia and Herzegovina.

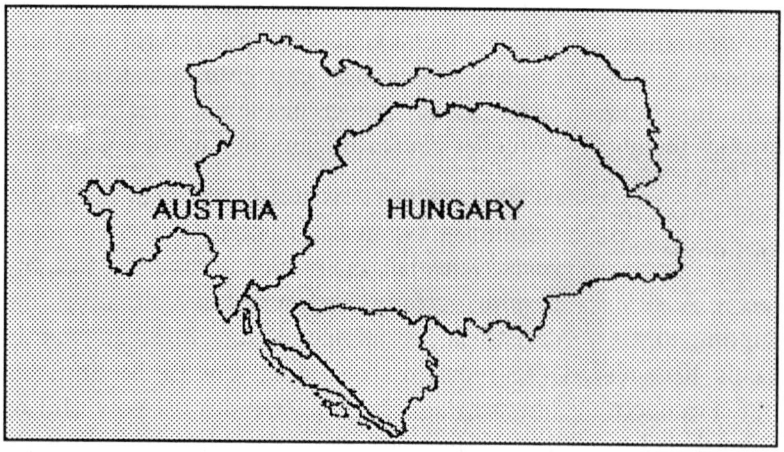

Figure 1-2 Map of Hungary in the Austro-Hungarian Empire

As a result of the compromise, Austria and Hungary each had its own parliament and government. The only unalterable link between the two was the person of the Emperor-King. Three ministries -- the Army, Foreign Affairs, and Finance -- were in common. The ministers alternated between an Austrian and a Hungarian, and each was responsible to a committee selected by the two parliaments. A series of renewable 10-year contracts dealt with a number of other important issues. Hungary's contribution to common expenses was set at 30% of her national income for the first 10-year period.

It was only on June 8, 1867 that Francis Joseph was crowned king of Hungary. Until then, as Emperor of Austria, he was not -- in Magyar minds -- possessor of the *Holy Crown* that represented the highest power in the Hungarian nation.

Hungary attained greater and greater power in the partnership over the decades. At some points there was great concern among Austrian politicians about the degree of control that Hungary exercised. This was especially the case when the 10-year contracts on compromise issues were being renegotiated.

In truth, Hungary's role in the Dual Monarchy continued to be limited by internal bickering over minor constitutional questions. An additional problem was created by the many nobles who were deprived of their privileges in 1848, but were largely unwilling or unable to do any productive work. They moved into civil service in large numbers and created a bureaucracy intent on self-perpetuation. Thus, members of the often German or Jewish middle class were barred from office, and the uneducated peasantry had no hope for advancement.

This situation led to, among other things, Hungary becoming a major source of immigrants to the United States in the late 1890's and early 1900's. While this emigration sapped Hungary's human resources of hundreds of thousands of working class Hungarians and Slovaks, conditions for most peasants in Hungary continued to be very poor.

Hungary's Jews were fully emancipated in 1867. Most Hungarian Jews were rather assimilated, taking Hungarian names and speaking the Hungarian language. The status of Jews in Hungary made it seem a "promised land" for Poland's Jews. A large immigration of Polish Jews to Hungary began concurrently with the rise in emigration of Hungarian peasants. The newly arrived and educated Jews largely entered commerce and the learned professions, since they were barred from civil service. By 1900, half of Hungary's medical doctors were Jews ... mostly unassimilated immigrants. This created yet another "minority" in Hungary, and created more problems in years to come.

The period of the dual monarchy was one of mixed blessings in Hungary. Major problems existed and others were hidden just beneath the surface. Yet, there was considerable progress. Industrialization continued, despite (or perhaps because of) much greater government interference in business than we are accustomed to in the western world.

Budapest became one of the great capital cities of the world. There, in 1896, Hungary's thousandth year of existence was celebrated. The *Millennium Exhibition* brought tourism, pride, and many permanent improvements to the city ... including the continent's first subway.

The liberal government that had negotiated the compromise remained in power for nearly forty years. But, in 1905 a series of radical governments gained power that resulted in ever-greater divisiveness in the nation. The competing factions were led by anti-intellectual ultra-nationalists and ultra-liberal intellectuals. The flames that ensued were fanned by a "gutter press."

To Hungary's credit, Count Stephen Tisza, who had previously served as premier, was called back to power. Tisza, faithful to the Compromise though that stance gained him much hatred, introduced several moderate reforms which in 1913 returned Hungary to sanity. Unfortunately, his opportunity to show his capabilities for peaceful development were ended when Francis Ferdinand, heir to the dual throne, was assassinated in Sarajevo on June 28, 1914. World War I -- the *Great War* -- began.

Francis Ferdinand was not mourned in Hungary. He was a pro-Slav, whose accession to the throne would have caused major constitutional problems in Hungary. But, his death ignited a war in which Hungary's role would be thankless. While Magyars fought, usually under Austrian or Czech generals, profiteering was rampant at home.

After ruling for 68 years, Franz Josef died on November 21, 1916. He was succeeded by his grand-nephew Karl (in English *Charles*), who on December 31, 1916 was crowned King Károly IV of Hungary. Figure 1-3 is the coronation photo of Karl with his queen Zita, and their only child, Prince Otto.

Obviously, Karl was able to command less respect in Hungary than Francis Joseph who had acquiesced to the Compromise, and who was revered by most of his subjects. The on-going decline in public morality was accentuated by the stress of war. On June 15, 1917 Tisza resigned. So, in Hungary, "things went from bad to worse."

Defeat of the Central Powers was inevitable. When the end came on the western front, Hungary's soldiers -- most of whom fought on the stalemated southern front and had not

Figure 1–3 The Coronation of Károly IV

suffered defeat on the battlefield -- returned to a country mired in confusion and virtual anarchy. The experiment of a federation of Danubian states -- which in fact was the scope of the Austro–Hungarian Empire -- had ended.

To digress for a moment, Karl died in 1923 after a short exile on the Portuguese island of Madiera. Queen Zita died only a few years ago at the age of 99. Otto von Habsburg, now a spry octogenarian and a German citizen, continues to play a significant political role in the formation of the European Community.

The Hungarian Postal Service

The independent Hungarian Postal Service was established as a part of the overall *Ausgleich* -- the Compromise of 1867. A provisional postal treaty was signed by the Emperor on April 21, 1867 and came into effect on May 1.

Effective June 1, a new issue of stamps was produced and used in Hungary. Because at the time there were no printing facilities in Hungary that could produce a quality product in the time required, the issue of 1867 was printed in Vienna. It was also used in the Austrian portion of the Dual Monarchy when supplies of the previous issue were exhausted.

These stamps have led to a recurring argument over which is the first stamp issue of Hungary -- this so–called "common" issue of 1867 or the issue of 1871, which was limited to use in Hungary alone. Rational arguments can be made that the issue of 1867 was a "forerunner of Hungarian stamps," a "common issue" of Austria and Hungary, or the "first Hungarian stamp issue." Fortunately, since this book is on aerophilately -- and in 1867 airmail lay far in the future -- I can avoid addressing that divisive issue here. Besides, more has probably been published on that aspect of Hungarian philately than any other. More, in this case, is not better.

In 1868, the area known as the "Military Border Districts" along the southern border of Hungary came under the full control of the Hungarian postal administration. In 1874, Hungary participated as an original founding member of the Universal Postal Union. Until then, mail from Hungary to other nations was forwarded in accordance with one–to–one treaties between Austria and other nations that were established before the Compromise.

Hungarian philately continued through the latter decades of the 19th century and into the early years of the 20th century in the conservative stamp issuing tradition that was typical of European nations of the time. Actually, the Hungarian postal administration's release of official picture postcards in conjunction with the *Millennium Exhibition* in 1896 was considered by some to be a rather radical action.

In the field we are focusing on here -- aerophilately -- Hungary was also somewhat radical and innovative. Hungary's issue of airmail stamps on July 4, 1918 made it only the fourth nation on earth to produce and use airmail stamps. It followed only Italy, which issued airmail stamps in 1917, and Austria and the United States which issued them earlier in 1918. As we will see, Hungary's first airmail stamps were issued for use on civilian airmail carried between Budapest and Vienna by military pilots for only three weeks.

Figure 1-4 Birth of the Hungarian Republic

In the aftermath of World War I, the turmoil in the country led to the use of Hungarian stamps in successor states, as well as many occupation issues that were massively forged. The inflation of the early 1920's also led to large issues of low-value stamps that continue to be virtually worthless today. In 1926, the currency reform led to a new stability in Hungarian stamp-issuing policy. Hungarian stamps of the late 1920's and the 1930's are noted for their fine design, limited quantities, and quality production. The airmail stamps of the period fall into that category and are discussed in the appropriate sections that follow.

Hungary Between the World Wars

Even before the conclusion of hostilities, a series of events commenced in Hungary that resulted in its loss of international respect and led to the harsh provisions of the Treaty of Trianon.

Turmoil. As the war's end approached, a basically undefeated Hungarian army collapsed due to loss of leadership in a nation suddenly divided by, among other things, the nationalistic ambitions of the non-Magyar peoples. On November 16, 1918 Hungary was declared a Republic under the leadership of Count Mihaly Károly. Figure 1-4 shows the special cancellation used on that day.

Károly is described by various authors as a "weak," "ineffectual," "well meaning," "liberal" "aristocrat." Clearly he was an enigma. Most certainly, as Sinor notes, "he was the last man able to realize" his aims. When on January 11, 1919 he was elected President by acclamation, he was little more than a figure-head. Two months of government by weakness and inaction allowed Czech and Romanian bands to move into much of the country.

On March 21, Károlyi somewhat unwillingly turned the government over to Béla Kun and a government of so-called "People's Commissars," who promptly declared Hungary to be a Soviet Republic. This alarmed the Allies, and French Gen. Smuts was sent to Budapest to try negotiations. Kun rejected the French ultimatum, and quickly was offered more favorable terms -- which he accepted.

Soon the Romanians went on the offensive, and faced only with the weak resistance of a disorganized Hungarian army, reached the Tizsa. At that point, Kun opted for pragmatism over political theory and asked Gen. Aurel Stromfeld of the former Imperial army to organize resistance. He was immediately successful in turning against the Czechs and in a few weeks drove them out of northern Hungary (i.e. Slovakia).

Despite these gains, the Kun government had little popular support and no foreign support was forthcoming. A Hungarian offensive against the Romanians in July failed and the Romanian counterattack caused the fall of the Communist government on July 31. On August 3, Romania captured Budapest.

The Communists had maintained an air force. They had eight squadrons of 6 to 10 airplanes. Each squadron was manned by about fifty pilots and support personnel. They also had a seaplane base at Csepel with 10-12 aircraft and a staff of 100. All these aircraft, as well as their support facilities, were taken or destroyed by the invading Romanians.

The fall of the Communist regime led to a few months of complete chaos. This period, usually called the *White Terror*, was punctuated by many acts of personal vengeance. For a very short time, Archduke Joseph, who had been appointed *Palatine of Hungary* by Charles IV, ruled as Regent. The anti-Habsburg French quickly forced his resignation.

Horthy. One of the Archduke's last acts was to appoint Admiral Nicholas Horthy as Commander-in-Chief of the *Hungarian National Army*. Horthy had been the last commander-in-chief of the Austro-Hungarian Navy and for a year had served as aide-de-camp to Emperor Francis Joseph.

Horthy was a cultured Anglophile with knowledge of several languages. Using his diplomatic skills much more than force, Horthy led his "army" into Budapest on November 16. By March 1, 1920 he was elected Regent of Hungary for 10 years, with powers similar to those of the King (except for the right to confer titles). This role was later given to him for life.

During Hungary's 16 months of turmoil, the peace conference in Paris was determining her fate. Sinor feels that Hungary would not have fared better if she was represented, due to "French obstinacy, American ignorance and British indifference." In any case, the result was that on June 4, 1920 Hungary was forced to sign the Treaty of Trianon giving up roughly two-thirds of her area and population.

Aviation. While from the perspective of the nation, the lost lands and the lost Magyar population were the all-consuming issue, from our perspective as Hungarian aerophilatelists, the aviation provisions of the treaty were also important. Hungary was forbidden to engage in any military aviation activities.

In January 1920, Horthy purchased eight airplanes in Vienna to establish a fledgling air force. When it became apparent what the treaty provisions would be, the airplanes were promptly turned over to a government-subsidized private company -- *Magyar Aero-forgalmi RT* or *MAEFORT*.

This attempt at subterfuge ended in a crisis a year later that resulted in the dissolution of that company and the destruction of all its assets. It was the first of a series of such subterfuges that put off the official establishment of a Hungarian air force until late in the 1930's.

Another aviation provision of the treaty forbade Hungarian companies to build any airplanes with military capabilities. In reality this precluded the manufacture of anything but small sport planes. The result of this was that -- according to the 1922 edition of Jane's -- the wartime airplane manufacturers were either out of business, or temporarily pursuing other interests. UFAG was functioning as a woodworking shop, and Lloyd was making furniture and upholstery for automobiles. MAG, the airplane engine manufacturer, was still making engines -- but, for tractors and autos.

During this early period, Stephen Petroczy -- a pioneer pilot who we will hear much more about later -- was Director of the Civil Aviation Section of the Ministry of Commerce. Most of the flying and other non-military aviation restrictions of the Treaty of Trianon expired in November 1922. This permitted the organization of *Magyar Légiforgalmi RT* (*MALÉRT*) and the beginnings of a domestic aviation and air transport industry. Despite this, throughout most of the 1920's only a handful of small sport planes were manufactured in Hungary.

In 1928, one of Hungary's biggest manufacturing companies -- the *Manfred Weiss Works* -- opened an aircraft and engine subsidiary on the island of Csepel, in the Danube just south of Budapest. They built Fokker airplanes and Bristol engines under license. In 1931, they built the first airplane of their own design -- the W.M. 10 -- a light, two-seat biplane for elementary training and

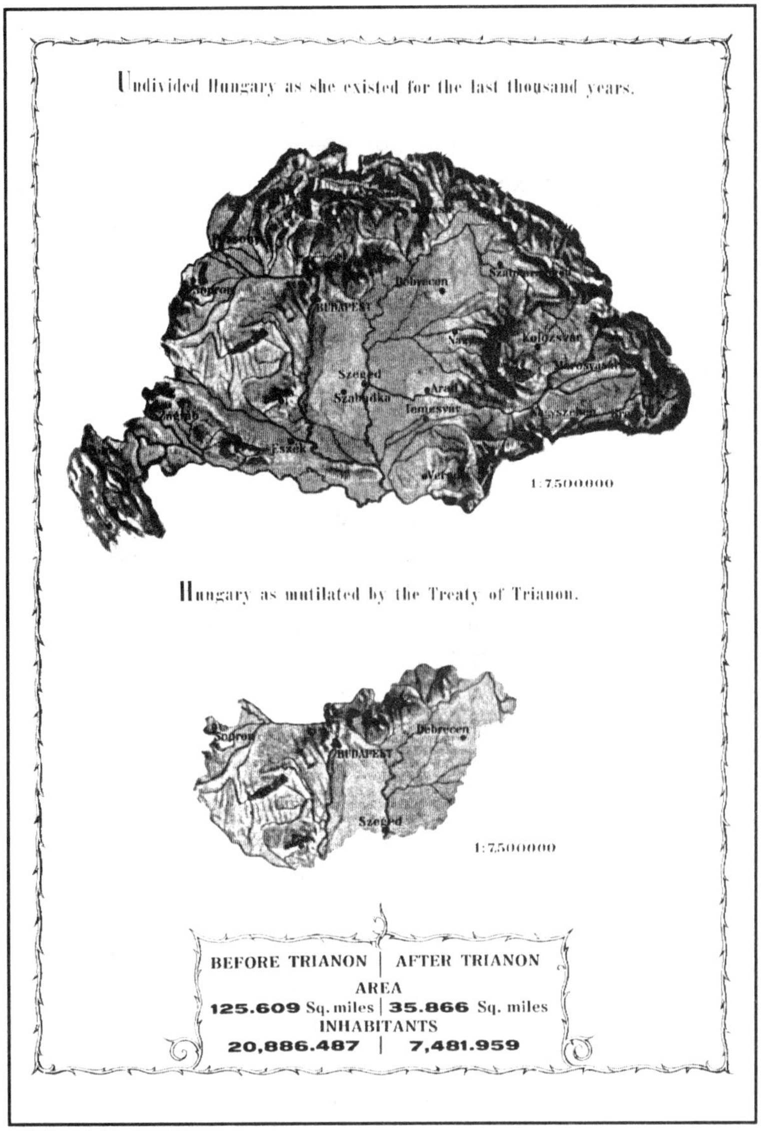

Figure 1-5 The Results of the Treaty of Trianon

sport flying. In 1935 they began producing an air-cooled four-cylinder in-line aviation engine of their own design. Throughout the 1930's, Weiss was the only commercial Hungarian aircraft company. [A few small sport planes also continued to be built by the *Sport Flying Association* of the Technical University of Budapest.]

Foreign Policy. The singular goal of Hungarian foreign policy between the world wars was the recovery of some of the lost territories. The nation's cry when it came to the acceptance of the Treaty of Trianon was "Nem, Nem, Soha!" -- *No, No, Never!* Figure 1-5 shows one of the many Hungarian postcards that flooded the world during the 1920's trying to expose the injustice of the treaty and the need for "Justice for Hungary."

Though most western statesmen recognized that Trianon was unsatisfactory, no one was ready to do anything about it. The French remained under the influence of their Czech and Romanian friends. The British were mildly sympathetic but were not in a position to do much. The first European political leader to realize the value of a stronger Hungary was Mussolini. He signed a friendship treaty with Hungary's premier, Count Stephen Bethlen, in 1927. Hungary's relationships with its traditional friends -- Poland and Turkey -- were good. Relationships with Germany and Austria were cool, but acceptable.

When Gyula Gömbös was appointed premier in 1932, he moved the country to the right. Gömbös is variously described as a "ridiculous personage" and "a man of great vision with a deep sense of social justice." Horthy himself referred to Gömbös as "colorful." Gömbös aimed his foreign policy at bringing together a Berlin-Rome "axis" -- a word he coined. In that he was successful, but, the result was greater economic dependence on Germany and the beginnings of Nazi ideology taking hold in Hungary.

The Economy. The period from the end of the First World War until mid-1924 was one of great economic instability in Hungary. These problems resulted from the political turmoil of the immediate post-war years. Of the defeated Central Powers, Hungary's inflation was second only to that of Germany in its impact on the nation and its people.

The war, the dissolution of the Austro–Hungarian Empire, the succession of ineffective radical governments that followed, anarchy, foreign occupation, the terms of the Treaty of Trianon, and the impact of the refugees from the "successor states" were the causes of the problem. All these combined to produce a *decrease* in the available capital resources of Hungary of over 95% between 1910 and 1921.

In March of 1921, Charles IV attempted to regain his throne. He got to Budapest, but withdrew when the allies threatened armed intervention. Shortly thereafter, Horthy appointed Count Stephen Bethlen premier. He deserves much of the credit for Hungary's recovery during the 1920's. But, before any progress was to be made, Charles made another attempt that was thwarted by armed resistance and a personal confrontation with Horthy. On December 3, 1921, Parliament dethroned the house of Habsburg for the third and last time.

Also in late 1921, an armed uprising against ceding western Hungary to Austria was undertaken by former army officers. The allies agreed in the October 12 *Protocol of Venice* to allow a plebiscite in Sopron and surrounding villages. The vote was overwhelmingly in favor of Hungary, and this city became the only lost territory returned to Hungary until 1938.

Bethlen, recognizing that Hungary needed substantial foreign capital to get back on her feet, applied for membership in the League of Nations. What had been until that time essentially a "victors' club," accepted Hungary in September 1922. In March 1924, the League approved a reconstruction loan to Hungary, over the objections of Czechoslovakia, Yugoslavia, and Romania.

It was only after Horthy's stabilizing role as Regent took hold, after the threat of re-establishing the Habsburgs was eliminated, and after the Allies were willing to allow considerable investment in Hungary, that the currency stabilized and the economy began to improve. The rate of inflation peaked in 1923, and the currency began to stabilize in early 1924.

The inflation was finally laid to rest when a new gold–based Pengó/fillér currency was introduced in July 1925. The first regular postage stamps in the new currency were issued on March 26, 1926. The Korona–valued stamps were valid until March 31, 1927. From an aerophilatelic perspective, the transition to the new currency took over two years – until the airmail stamps of 1927, denominated in the new Pengó/fillér currency, were issued on September 1.

Hungary's economic recovery was so profound that it encouraged major foreign capital investment in both manufacturing and agriculture. Industrial and farm production rose rapidly and foreign trade was brisk. The worldwide depression that started in 1929 put an end to that prosperity. A small country could not isolate itself from the world. A government financial crisis in 1931 led to Bethlen's resignation. It was more than two years before modest prosperity again returned to Hungary.

The Returned Territories. The *Anschluss* of 1938 that joined Austria to Germany produced a new situation. The Reich was now on the borders of Hungary and it appeared after the *Agreement at Munich* that the world would let aggression go unpunished. The rightward movement in Hungary continued with the appointment of Imrédy as premier, including anti–Semitic laws (though "mild" compared to those of Germany).

Munich set a process in motion that allowed Hungary to recover some of her lost territories. In the First Vienna Award of November 2, 1938 Germany and Italy awarded Hungary a narrow strip of southern Slovakia with a population of about 800,000. This small step satisfied no one. When in March of 1939 Germany seized western Czechoslovakia, Hungary marched into easternmost Czechoslovakia -- known philatelicly as Carpatho–Ukraine, but called by the Magyars *Ruthenia* -- and took it without opposition.

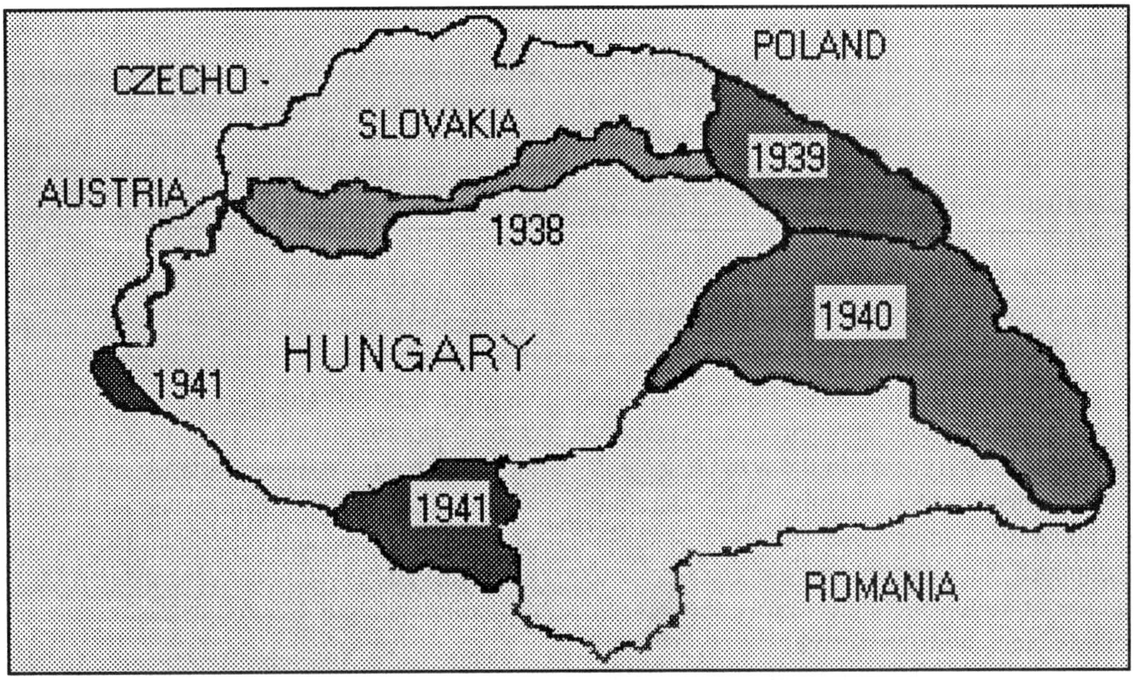

Figure 1-6 Trianon Hungary and the Returned Territories

This gave Hungary a border with Poland, which soon had important implications. Minor border changes were also negotiated with the Nazi–puppet regime of Slovakia.

Figure 1–6 shows a map of Hungary as it existed between the World Wars ... that is *Trianon Hungary*. The context of the map is historic Hungary (less Croatia). The highlighted areas were "awarded" to Hungary or annexed by Hungary in the 1938–41 period.

The Axis Powers in 1940 made a "Second Vienna Award" to Hungary. This time it was northern Transylvania that was detached from the Trianon borders of Romania. Finally, after the German invasion of Yugoslavia in 1941, Hungary annexed a part of the Serbian Banat, and a small portion of Slovenia that had been part of Hungary's pre–WWI *Vas megye*.

These awards and annexations led to two little-known airmail routes that are discussed in the chapter on *Airmail and the Returned Territories*. As a part of Hungary, the "returned territories" were to last only until the war's end. Trianon Hungary is essentially the Hungary of today, with further minor concessions to the Czechs at the "Danube bridgehead."

The Coming of War. In February 1939, a man of great moral rectitude, Count Pál Teleki, took office as premier.

When war broke out, Hungary's position was extremely precarious. It was apparent that a German victory would result in the absorption of Hungary into the Reich. Those who anticipated a German victory wanted to ensure friendship with their future ruler.

Neither Horthy nor Teleki were in this camp. They believed the English–speaking maritime nations -- Great Britain and the United States -- would stick together and ultimately triumph. They wanted, at all cost, to keep Hungary out of the conflict. Hungary refused to have anything to do with the attack on Poland, and gave refuge to many Polish military officers who escaped over the common border. Most of these Poles eventually made their way to England, to continue their fight from exile.

The fall of France did not change the Hungarian government's non–belligerent attitude; though, the pro–German forces within Hungary were strengthened by that fact. In April 1941, Hungary became the unwilling accomplice of Germany's attack on Yugoslavia.

Teleki was unable to prevent the German army's transit of Hungary, and committed suicide. Teleki's successor, Bárdossy, was less reluctant to cooperate with Germany and yielded to German pressure to declare war against the Soviet Union on June 27, 1941. By December, Hungary was also officially at war with Britain and the United States. Though Hungary was in the war, she was still reluctant to commit significant resources.

Hitler did not allow such freedom to his reluctant ally for long, and demanded the active participation of Hungary in the war. Hungary's 200,000 man Second Army, with the best equipment the nation possessed, was dispatched to support the Germans in the Ukraine. In January of 1943, the Soviets broke through the German lines and the Second Army, assigned to rearguard action to protect the retreating Germans, was annihilated. Hungary's ability to be a meaningful military participant in the war ended. On March 17, 1943 Horthy was summoned to Germany to visit with Hitler. During his absence, the German army occupied Hungary. Nazi forces in Hungary -- the German army and their Hungarian admirers -- now had the upper hand.

With little real power left on his return, Horthy concentrated, with some success, to slow the deportation of Hungarian Jews, and persisted with his efforts to extricate Hungary from the war. Finally, in October 1944, he made a radio broadcast unilaterally declaring an armistice. He was promptly arrested by the Gestapo and carried off to Germany. By this time, of course, the Soviet Army had moved into Hungary.

On Christmas Eve of 1944 the Red army reached Budapest. Through the obstinance of the retreating Germans and the aggressiveness of the invading Russians, Budapest and much of the rest of Hungary were destroyed by April of 1945 when all fighting finally ended.

Recent History of Hungary

Soviet Occupation. The Soviet High Command established a provisional government in Debrecen on December 23, 1944. This government was led by General Béla Miklós, who had surrendered his army to the Soviets two months earlier. Four "approved" parties were permitted a part in this government, with the Communists firmly in control. Since the Communist party had been banned under Horthy, its leadership was made up primarily of emigres who went to the Soviet Union after the fall of the Kun government in 1919.

The war was still going on in the western part of the country. In their retreat from Hungary, the Germans took most of the nation's portable wealth; but, at least spared personal possessions. The invading Soviets had no such scruples. They systematically raped and looted the nation with great thoroughness.

On January 20, 1945 the Hungarian provisional government signed an armistice with the allies under which Hungarian civil authority reverted to the Trianon boundaries. In November of 1945 elections were held. Despite Soviet efforts to screen candidates and disenfranchise voters, the Smallholders Party -- which favored establishing a western–style democracy -- won an overwhelming mandate with 60% of the vote. But, reality set in.

The nation was under Soviet military occupation, and all government decisions were subject to the approval of the Allied Control Commission. The ACC was dominated by the Soviets due to American and British focus on maintaining control in the areas of Germany and Austria they occupied. Therefore, the Soviets -- while permitting Zoltán Tildy, the leader of the Smallholders, to be named premier -- kept effective control of the government by appointing Communists to lead all the key ministries. This included the Ministry of the Interior, which controlled the police.

On February 1, 1946, Hungary was proclaimed a republic with Tildy as president. Until the formal signing of a peace treaty on February 10, 1947, the real power in Hungary continued to be Georgi J. Pushkin, the Soviet member of the Allied Control Commission.

Inflation. The 19 months from January 1945 through July of 1946 in Hungary represented the greatest inflationary period ever to befall any nation on earth. The inflation had many causes ... the loss of the war ... the devastation caused by the retreating Germans and invading Soviets ... the burden of a military occupation. But, most of all, it was the result of one of the first east-west post-war confrontations. The U.S. controlled all of Hungary's assets that were outside the country, and would not return them to Hungarian control for fear they'd fall into Soviet hands. On the other hand, the Soviets demanded war reparations that far exceeded Hungary's total production capacity, to legitimize their claim on Hungary's overseas assets. The stalemate went on while Hungary essentially reverted to a barter economy.

The monetary crisis only ended when both sides backed down. The Soviet Union scaled back its reparations demands, and the U.S. released most of the Hungarian assets they controlled. On August 1, 1946 a new fillér/Forint currency was established. Though substantially devalued over the years, and still not a medium of free exchange, the Forint continues as Hungary's currency today.

The philatelic tale of Hungary's monstrous inflation is best seen in the regular postage of the time. A brief view of it can be seen in the very rare airmail of May-July 1946 that is discussed in the final chapter of this book.

Aviation. When the war ended, not a single aircraft or piece of aviation support equipment survived. Budapest's airport was in ruins, as were all the smaller airports in provincial cities. At the insistence of the Soviets, no civil aviation was permitted to enter or leave the country until the peace treaty was signed. In 1946, Pan Am was refused permission to extend its route to Budapest. As we shall see, international airmail was carried into and out of Hungary by surface means during this period.

On March 29, 1946 Hungary and the Soviet Union signed an air transport agreement. This led to the formation of the Magyar-Soviet Airways Company -- known as *Maszovlet*. This 50-50 operation was initiated using aircraft and support equipment provided by the Soviets. The initial aircraft were DC-3's built in the Soviet Union during the war. The first domestic operations began on October 15, 1946.

By 1948, several foreign airlines were servicing Budapest and *Maszovlet* was flying to other eastern-bloc capitals. In 1954, the Soviet Union ceded its interest in *Maszovlet*, and it became the Hungarian national airline *Malév* that continues in operation today.

After the Peace Treaty. In 1945, the allies only addressed the question of peace treaties with the minor players in the Axis coalition, including Hungary, but not Germany or Japan. A peace treaty was signed between Hungary and the allies in Paris on February 10, 1947. Hungary had virtually no input into the process. Those interested in the detailed history of this period should read Stephen D. Kertesz' 1992 **Between Russia and the West**.

Philip E. Mosely, a member of the American delegation at the conference, writes that the Hungarian government assumed "the issues would be treated on their merits by the [allies who were] concerned primarily with promoting peace and stability in the Danubian region." He notes this was not the case and that: "The struggle over the formal terms of the treaty was merely one aspect of a more general struggle to extend or confine Soviet power in Europe. In that struggle Hungary had little to hope for and much to lose."

Hungary expected the withdrawal of Soviet troops with the signing of the treaty. That didn't happen. The Soviets chose to remain on the pretext that they needed to maintain communication lines with the occupation forces in the Soviet Zone of Austria. As we know now, it was to be the early 1990's before Soviet troops were completely gone from Hungary.

The Soviets also continued to interfere with government operations in Hungary after the peace treaty was signed. Non-communist members of parliament were arrested at Soviet direction whenever it suited the communists' purpose. Finally, on August 31, 1947, rigged elections were held which ousted the Smallholder's party and initiated over forty years of Soviet–line one–party rule.

We are all familiar with the abortive Hungarian Revolution of October 1956 which left the world wondering if more should have been done to support Hungary's patriots. The more recent events, in which Hungary played a pivotal role in the fall of the *iron curtain*, are also familiar to each of us. Because these events all come after the period which is the focus of this book, I will end this historical overview here.

By some accounts the pioneer period of Hungarian airmail ended in 1918 with the end of World War I. Others might argue for 1927 when an integrated airline system covered Europe, or 1939 when Hitler brought the world a new concept of war. There is no doubt however, that the pioneer period did not extend beyond 1946 when the recovery from World War II began. Therefore, giving the subject the greatest possible latitude, I have selected 1946 as the date to end this book. Let's now move on to the real purpose of this book ... a study of *The Pioneer Period of Hungarian Airmail*.

The Beginnings of Airmail in Hungary

Anything that you might call a "regular" airmail service did not come to Hungary until World War I. As in other parts of the developed world, the two decades before the war were the period when Hungarians really first took to the skies. In doing that, they sometimes took mail with them, and certainly used the mails to describe their experiences.

Obviously, the development of aviation in Hungary -- particularly in the military sector -- was inextricably tied to developments in Austria. Therefore, whenever appropriate, Austrian aeronautical history is also introduced. But, the aerophilatelic material discussed always has a strong Hungarian connection. In the early years of aviation, balloons, dirigibles, and airplanes were all a part of the story.

Flight: from Myth to Reality

I will not claim that Icarus was a Hungarian! Every nation has among its heroes those who unsuccessfully attempted flight, and those who were the true aviation pioneers in their corner of the world. Hungary was no exception. After briefly mentioning some of the tales of early Hungarian attempts at flight, I will discuss three of the Hungarian pioneers of aviation. Hopefully this will establish Hungary as a nation with a serious history of aeronautical achievement -- and lead us into the aerophilatelic aspects of this book.

Figure 2–1 The 1967 Verancsics Stamp

The first recorded Hungarian aviation pioneer was Faustus Verancsics (1551–1617), the Bishop of Csanád, who was a person of considerable literary and technical accomplishment. He spent the latter years of his life in Italian monasteries pursuing scientific questions. In 1616, the book **Machinae Novae** was published in Venice. It pictured among its technical innovations a parachute apparatus attributed to Verancsics. Figure 2–1 shows a Hungarian airmail stamp of 1967 picturing the Verancsics parachute device.

The earliest known flight experiments within Hungary itself may have occurred in the county of Nyitra (now Slovakia) in 1760. Contemporary drawings show Jaisge Cyprian, a surgeon in the Red Monastery, attempting flight using wing–like structures attached to his arms. Fortunately, his experiments concluded on the orders of the Bishop before he duplicated the fate of Icarus -- whose mythological attempt to fly used the same approach and ended in death.

The first balloon flight in Hungary took place in 1794, about a decade after the historic Montgolfier first flight in Paris. Professor Stephen Szablik made a brief free balloon flight entirely within the confines of Pest. The newspaper **Magyar Merkurius** ("*Hungarian Mercury*") carried a picture of Szablik's balloon on October 28 of that year.

During the 19th century, balloons appeared over Hungary with increasing frequency. These early flights, of course, brought stories of how people reacted to balloons. One example tells of a captive balloon that broke away from its moorings at the 1873 Vienna World's Fair. Eventually it landed at Moson in Hungary. There, the story goes, a farmer attacked the "monster" with a pitchfork and stabbed it to death. Since no mail was aboard, I guess this flight lacks serious philatelic credentials.

Three years earlier, many carefully documented mail–carrying balloons left Paris during the Prussian siege. Though none of these Parisian balloons landed in Hungary, a few carried mail that was addressed to, and eventually delivered in, Hungary. These interesting pieces are unquestionably the first airmail handled by the recently independent Hungarian postal administration. The small envelope shown in Figure 2–2 was carried aboard the balloon *L'Archimede* from Paris to Holland on November 21, 1870. The cover went by normal posts from Holland to Pest via Vienna, where a transit mark was applied. These so–called *Ballon Montes* dispatched to central and eastern Europe are quite scarce.

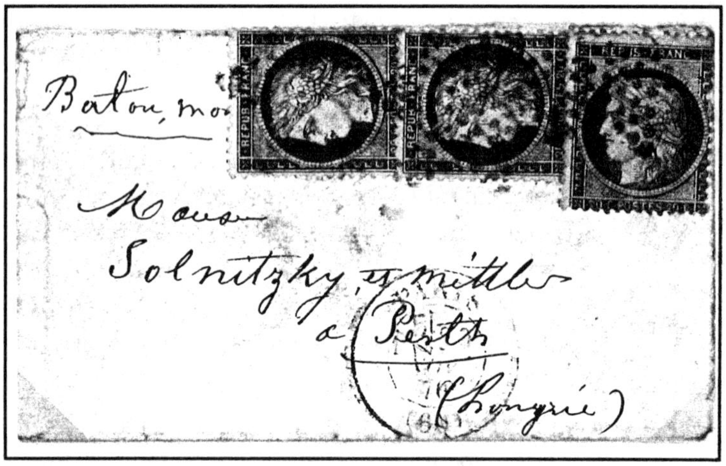

Figure 2–2 1870 Paris Balloonpost to Hungary

Louis Martin, a professor at Kolozsvár University (today Cluj, Romania), in 1893 showed with a flying model that copying the wing motion of birds could not solve the problem of heavier–than–air flight. His death in 1897 prematurely ended his attempts at practical, controlled, heavier–than–air flight.

Pioneers of Hungarian Aviation

The unique political relationship that existed between Austria and Hungary during the early period of aviation successes sometimes blurs national distinctions. So, I will try to focus on aviation pioneers who were born in Hungary -- even if they worked in Austria or elsewhere. When attempting to introduce the pioneers of Hungarian aviation, I will almost certainly omit someone's favorite hero. In every list of pioneer aviators, there are some who *have* largely been forgotten, some who *should* be forgotten, and a few who *never* will be forgotten. I have chosen to focus here on one pioneer of Hungarian aviation in each of these three categories. Of course, many additional Hungarian aviation pioneers are mentioned along the way in this and other chapters.

Dávid Schwarz, a Hungarian born in Keszthely in 1845, after a successful career as a timber merchant in Zágráb, invented the rigid airship. The construction of this type of airship became practical with the great *decrease* in the price of aluminum in the 1880s, which resulted from a new extraction process. After a review of his plans, the Austro–Hungarian army decided not to support Schwarz's work.

He went to St. Petersburg to build his airship with Russian support. This 1893 attempt failed due to the poor quality of the materials. In 1895, the German government agreed to pay Schwarz 300,000 Marks when an airship of an improved design made a successful ascent and landing. The airship's construction was partially financed by the German patent attorney and aluminum manufacturer, Carl Berg. Schwarz suffered a heart attack in January of 1897, on the very day he received the invitation to make a test flight in Berlin.

Figure 2–3 The Schwarz Airship Stamp

Figure 2–3 shows the Schwarz airship as pictured on a 1967 Hungarian airmail stamp.

Figure 2–4 Schwarz's Airship Preparing for Takeoff in Berlin

Schwarz's widow, Melanie, helped prepare for the test flight. This flight took place at Templehof field in Berlin on November 3, 1897. The 175 foot long airship was constructed of aluminum sheeting, less than 1/100th of an inch thick, over an aluminum framework. 130,500 cubic feet of hydrogen provided the airship's lift. A 16 hp gasoline engine and two belt–driven propellers provided forward motion.

A German soldier named Platz volunteered to pilot the airship on its test flight. Some time after a successful lift off, the propeller belts slipped from their pulleys. Attempting to correct the situation, the inexperienced pilot released too much gas through the valve and bounced to a crash landing ... destroying the airship. He saved his life by jumping clear just prior to the impact. Figure 2–4 shows a photo of the Schwarz airship at Templehof preparing for takeoff.

Count Ferdinand von Zeppelin witnessed this test flight and crash. In early 1898, Zeppelin (through intermediaries) acquired all rights to Schwarz's patents and ideas from the widow for 15,000 Marks. Berg, who also helped finance Zeppelin's work, arranged the transaction.

Only two years later, the first of Zeppelin's rigid airships lifted off on its first flight from Lake Constance. His airships, which made the name *Zeppelin* known throughout the world, were based partially on Schwarz's technology. Already, Dávid Schwarz -- one of the great pioneers of Hungarian aviation -- was all but forgotten.

In thinking about pioneer aviators, this question always comes up: "Who was the first to fly an airplane in Hungary?" A Hungarian paper manufacturer from Arad, named **Sándor Némethy** claimed this honor. His airplane -- which had the appearance of a bicycle with wings -- supposedly made a "hop" in early 1909. Most experts do not accept his claim as a true flight. Figure 2–5 shows a photo of Némethy's *Flugrad* on an advertising postcard of the period. (*Flugrad* is German for "flying bicycle.")

Némethy is sometimes mistakenly recognized as an important pioneer aviator. This is perhaps out of excessive nationalistic pride. Despite later experiments with a second aircraft, Némethy's work never did contribute in any way to the mainstream of the development of aviation in Hungary.

Figure 2–5 Némethy's "Flugrad"

Born in Budapest in 1881, **Tódor Kármán** (known in the U.S. as Theodore von Kármán) was a son of a renowned university professor. His mother was a cultural leader of Hungary. Kármán received a Master of Engineering degree from the Royal Technical University of Budapest in 1902. He spent a year in the army, as required at the time, and left as a First Lieutenant in the reserve 61st Heavy Artillery Regiment, with which he performed his annual training until the war.

After doing research in internal combustion engines in Hungary, Kármán resumed his studies at the University of Göttingen in Germany. There he received his PhD in 1909. After three years of teaching at Göttingen, he assumed the position of professor of aeronautical engineering and director of the Aeronautical Institute at the University of Aachen. He held this position, at least nominally, until 1930. Kármán's auto-biography, **The Wind and Beyond**, is an excellent source of information on his most interesting and productive life.

With the onset of war in 1914, Kármán was called back to his army unit. When the military became aware of his scientific credentials, Kármán notes that: "They reassigned me to the job of determining the type of artillery needed for the defense of Budapest in the event of a Russian siege. I arrived at work every morning about 9 AM, and the first thing I did was to check the newspapers to discover whether the Russians were any closer to Budapest. They usually weren't, so for the rest of the day I could relax."

In 1915 he joined the Austro-Hungarian air force headquarters in Vienna. He believed that the "air force of that period was a joke," with only about 50 ill-trained pilots and a few poorly maintained airplanes. Shortly thereafter he established, and was made chief of, an experimental aeronautical laboratory at Fischamend, outside Vienna.

Among the projects Kármán led at Fischamend was an adaptation of the German mechanism (invented by Fokker) which synchronized a machine gun for firing through an airplane's rotating propeller. Also at Fischamend, Lt. Louis Bencze developed a plastic cover with safety valve to keep aircraft fuel tanks from exploding when pierced by bullets. A similar independent invention was made in the U.S. at about the same time.

Figure 2–6 The PKZ–2 Helicopter Stamp

During World War I, tethered "kite balloons" were used for *artillery spotting*. The crew of the balloon radioed information to the ground forces. Balloons were very expensive and difficult to support, and more importantly made very large targets. The Austro–Hungarian military recognized the helicopter as presenting a much smaller target for enemy gunners. Also, it was not as flammable as a hydrogen–filled balloon and represented a viable air defense platform. Kármán estimated that an artillery spotting unit could be *reduced by* four wagons and 110 soldiers by simply replacing the balloon with a tethered helicopter.

Col. István Petróczy directed the project to develop such a helicopter. A pioneer pilot himself, Petróczy was the chief instructor for Army balloon crews. He died in 1957. While Petróczy managed the project, he did not make any technical contributions to the helicopter.

Beginning in 1917, Kármán focused all his work on the pioneering helicopter experiments. Two fine aeronautical designers, Zuravec and Asbóth, assisted him in this project. The helicopters were designated the PKZ-type for Petróczy-Kármán-Zuravec. It is notable that all four developers of these helicopters were Hungarians. Before building a full-sized aircraft, Kármán performed laboratory tests on small elastic-driven models and then a 65 lb. model powered by a 5 hp compressed air engine.

The project engineer for the PKZ helicopters was Vilmos Zurovec. His main responsibility was developing the power source. A 200 hp electric motor powered the first model, the PKZ-1. Though it was capable of lifting off with the weight of three people, it was unsuccessful in achieving altitude before the engine overheated. As a result, the PKZ-1 approach was abandoned. Figure 2-7 shows a photo of the PKZ-1 attempting takeoff.

Figure 2-7 The PKZ-1 Attempting Takeoff

Its successor, the PKZ-2 weighed 4,500 lbs. and was powered by three *LeRhone* 120 hp rotary gasoline engines. The PKZ-2 had three "arms." Each held an engine. The arms also served as attachment points for the mooring cables. Two counter-rotating propellers, operating at 600 rpm, provided the helicopter's lift.

These rotors were *below* the crew's cabin which Kármán described as "an open metal can." The crew consisted of an observer and a gunner. From his position in the can, the observer was to communicate with the ground by radio. The crew also had a large parachute device that was capable of lowering the entire aircraft in the event of equipment failure. Figure 2-6 shows the PKZ-2 as illustrated on a 1967 Hungarian airmail stamp. The balloon-like balls attached to the base and the arms are just that ... balloons ("pneumatic buffers") to cushion a hard landing.

The wooden rotors were 20 feet in diameter, and were designed by Oszkár Asbóth, who during the war was director of the state propeller factory. Later in life, he became well known for his development of helicopters and autogyros in Hungary in the late 1920's.

Many of the test flights were unmanned. A sandbag in the crew compartment simulated the crew's weight. Kármán said he flew in the helicopter, on at least one occasion, for about one hour at an altitude of about 100 feet. He noted "the strong impression that life as an observer in [the PKZ-2] could be very uncomfortable."

Thirty ascents were made at the test site near Budapest. During one of those flights, an altitude of 130 feet was attained. On June 10, 1918, the PKZ-2 helicopter became unstable in flight and was severely damaged in a crash landing. The PKZ helicopters had stability problems whenever the cables went slack, often due to an unanticipated wind shift. This happened that day, precipitating the unfortunate crash, with many dignitaries looking on. A contract was awarded to the Lipták factory in Budapest to produce PKZ-2 helicopters. Their manufacture never began due to the approaching end of the war.

The PKZ helicopters were conceptually similar to the experimental ABC (Advancing Blade Concept) helicopter developed by Sikorsky Aircraft in the 1970's. This author was responsible for Sikorsky's flight test data processing during the experimental testing of the ABC aircraft.

Figure 2-8 is a photo of the PKZ-2 helicopter in tethered flight near Budapest.

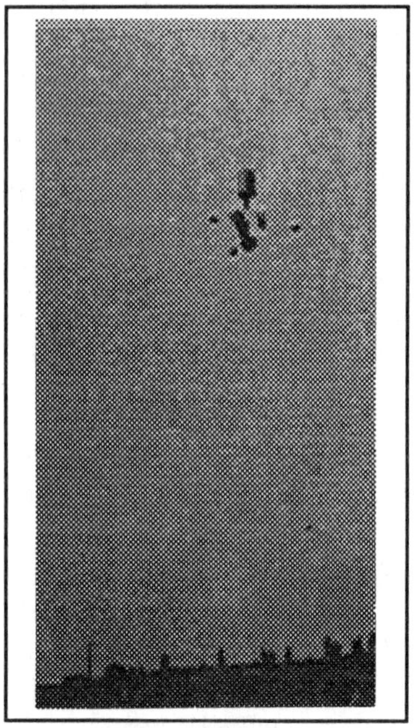

Figure 2-8 The PKZ-2 Helicopter in Flight

After the war ended, Kármán wanted to return to teaching in Germany, but this was impossible for a time, since the Allies did not allow military officers of defeated countries to leave their homeland. Near the end of 1918, he joined the Educational Council of Count Michael Károly's government. His task was the reorganization and modernization of the University of Budapest.

With the coming of the communist government of Béla Kun in 1919, he stayed on as Undersecretary of Universities. At the end of Kun's four months in power, Kármán says "I wasn't sure how any member of the Kun government would fare, no matter what his political persuasion, so I played it safe ... and made preparations to go to Aachen."

During the 1920s, continuing on the faculty at Aachen, he established extensive international contacts and lectured in Asia and America. In 1930, Kármán accepted appointment as director of the Guggenheim Aeronautics Laboratory at the California Institute of Technology. He became a U.S. citizen in 1936. In addition to founding the Aerojet General Corporation and the Jet Propulsion Laboratory, he was among the most important advisors to the U.S. Air Force before, during, and after World War II.

Figure 2-9 U.S. von Kármán Stamp

In February 1963, President John F. Kennedy awarded Theodore von Kármán the first National Medal of Science. Kármán died just three months later. In summarizing his career, Kármán wrote the following. "I am an optimist. I believe in the goodness of the future. And if I have done a small bit to help bring it about I am content." This great Hungarian-American pioneer of aviation lived and was productive from the *pioneer period* into the *jet age*, and even to the beginnings of the *space age*. The world will long remember the legacy of scientific accomplishment left by Theodore von Kármán. Figure 2-9 shows the 1992 U.S. stamp honoring him.

Pre-World War I Aviation

This chapter provides an overview of pre-WWI aviation in Hungary. It is organized by type of aircraft ... balloons, dirigibles, and airplanes in that order. The major topics of interest to aerophilatelists: the *Army Balloon Mail* and the *1910 International Flying Competition* are covered in separate chapters.

There was one "airmail event" of 1900 that did not involve aircraft ... but rather pigeons. An international picture postcard exhibition was held in Budapest in June. Special cancellations commemorated the exhibit. A pigeon-post vignette is known, but no flown messages have ever been reported.

Balloons. Balloon experiments and exhibits had occurred occasionally in Hungary during the last half of the 19th century. Popular interest in ballooning reached a peak during Hungary's *Millennium Exhibition* in Budapest in 1896 when the Frenchman Capt. Louis Godard took exhibit visitors aloft in a captive balloon.

A card shows Godard's balloon tethered above Budapest's City Park. He took people for "rides" in the balloon, but there is no evidence the cards were carried aloft. Boesman, an authority on balloon mail, claimed only three of these cards have survived. I believe that estimate is low -- but, the Godard cards are certainly extremely scarce. The example in Figure 2-10 was actually posted later, in 1899. A French version of the Godard card exists.

In 1902, a balloon flying exhibition was held in Budapest, providing unflown advertising cards as aerophilatelic memorabilia. A 1905 picture postcard shows the first flight of the Magyar Aero Club's balloon *Turul*.

Shortly after Hungary's millennium, the Austro-Hungarian Army began to make fairly regular balloon flights from Fischamend, Aspern, and Wiener-Neustadt in the Vienna area. The objective of these flights was a combination of training and weather research. Some of these flights ended in the Austrian and Bohemian provinces of the Empire -- and even in Germany. Many of the flights landed in Hungary due to the prevailing westerly winds. The chapter on *Army Balloon Mail of 1898-1914* covers this material.

Dirigibles. Dávid Schwarz, who was a Hungarian born in Keszthely in 1845, invented the rigid airship. Count Ferdinand von

Figure 2-10 The 1896 Godard Balloon Card

Figure 2-11 Maneuvers Card Picturing the LZ-3

Zeppelin witnessed a demonstration flight of his airship in Berlin. Schwarz's patents and ideas influenced the development of Zeppelin's airships. The chapter on *Pioneers of Hungarian Aviation* provides more information on Schwarz and his aluminum airship.

The earliest known Hungarian philatelic item relating to a dirigible was a souvenir postcard of the joint Austro-Hungarian-German maneuvers near Veszprém in 1908. It pictured a Zeppelin over the royal headquarters in that city, which commemorated the planned participation of the LZ-3 in those maneuvers. That Zeppelin was being rebuilt for the German army, but its reconstruction was completed too late for it to participate in the maneuvers.

Figure 2-11 shows the reverse of that card. [Note: This 1908 card is a part of the correspondence of balloon pilot Siegfried Heller. Heller's mail is a treasure-trove of Austro-Hungarian army aviation history. His widow, unfortunately, removed most of the stamps from the mail before offering it for sale after Heller's death.]

The Austro-Hungarian army accepted its first airship in December 1909. It was the non-rigid *Parseval PL-4*. A picture postcard showing it in flight over Pozsony, Hungary that same month exists. The military designation of this airship was the M-I. Its specifications were: length 50m, diameter 8.6m, gas volume 2,300 cubic meters,

Figure 2-13 The M-II at Its Base

top speed 45km/hour. It was built in Austria to the Parseval Type C design. It carried four passengers and enough ballast and fuel for a twelve hour flight. Its base was Fischamend, Austria. Figure 2-12 shows a photo of the M-I beside its hangar in Fischamend.

Figure 2-12 The M-I Beside Its Hangar

In 1910, the Lebaudy-Juillot-6, a semi-rigid airship was added to the Austro--Hungarian army's

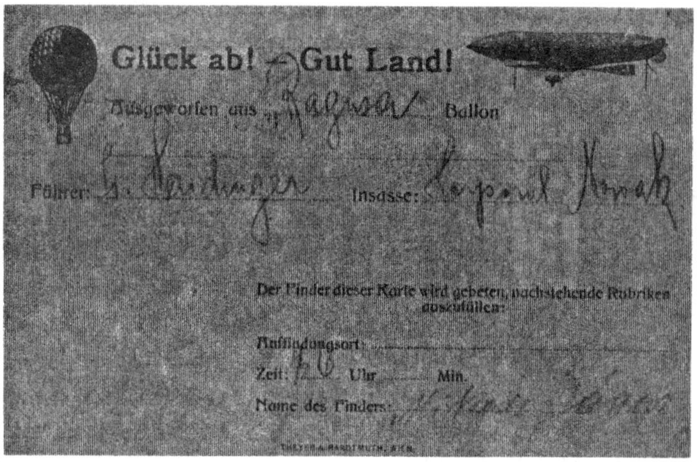

Figure 2–14 ***Ragusa*** **Drop Card Picturing the M–II**

fleet. It had the military designation M–II. Its specifications were: length 70m, diameter 11m, volume 4,800 cubic meters, top speed 45km/hour. The Austrian Daimler works built it to Lebaudy specifications. The M–II was stationed at Örkény, Hungary (50km southeast of Budapest).

The M–II was wrecked in March 1911, with all aboard lost. After repairs, it was again in operation by late 1912. Figure 2–13 is a photo of the M–II at its base.

Figure 2–14 shows a very unusual bright blue 1911 card dropped from the balloon *Ragusa*. The reverse of this card pictures both a balloon *and* the M–II dirigible.

The Körting–Wimpassing (K–W 1) non–rigid airship was completed in 1911 and given the military designation M–III. It was an adaptation of the Parseval type airships with: length 65m, diameter 10.5m, gas volume 3,600 cubic meters, top speed 49 km/hour. In June 1914, it was wrecked in a collision with an airplane. All the occupants of both the airplane and the M–III were killed.

Documented, flown mail carried by these airships is exceedingly rare. Since dirigibles provided controlled flight, drop cards did not serve a useful weather research purpose as they did when dropped from free balloons. There are only three known examples of dirigible drop cards (two from the M–I and one from the M–II airship). Unfortunately for Hungarian airmail collectors, the M–I cards were dropped over its base in Wiener–Neustadt, and the M–II card was found and posted at Linz, Austria. Figure 2–15 shows both sides of a 1911 postcard picturing the M–I. This card was carried on a flight of the airship to Hungary, and posted on landing by Lt. Tauber.

Figure 2–15 Landing Greetings Card of the M–I

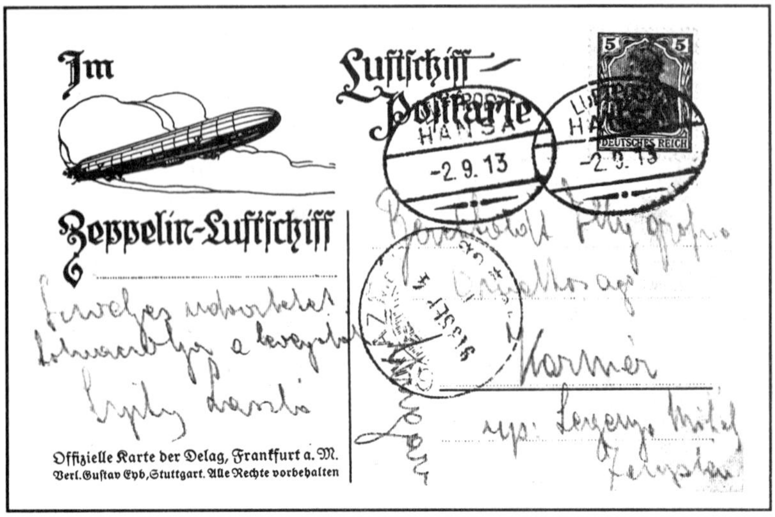

Figure 2-16 *Hansa* Card Destined for Hungary

Between 1911 and 1914, Zeppelins were in regular use in Germany for carrying both passengers and mail. These airships were the *Deutschland*, *Schwaben*, *Viktoria Luise*, *Hansa*, and *Sachsen*. Occasionally mail flown on these Zeppelins was addressed to Hungary. After being flown in Germany, such mail was carried to its destination in Hungary via normal surface means. Figure 2-16 shows a card flown by the *Hansa* destined for Nagykázmér, Hungary.

In 1913, a Zeppelin to be designated the M-IV was ordered by Austria-Hungary. Delivery was not made before the beginning of World War I, and so the military of the Dual Monarchy never operated a Zeppelin airship. But, German Zeppelins operated within Hungary during the war. See the chapter on the *Zeppelin Base near Temesvár*.

Airplanes. As mentioned in the chapter on the *Pioneers of Hungarian Aviation*, Sandor Némethy claimed to be the first to fly in Hungary. He claimed his airplane -- which had the appearance of a bicycle with wings -- made a "hop" in early 1909. This claim is normally discredited as a legitimate flight.

The famous French aviator, Louis Bleriot, must be credited with piloting the first airplane in Hungary. He performed this fete on October 17, 1909 at Rákosmezö, outside Budapest.

Some believe that Bleriot chose to honor Hungary with his presence due to the influence of Vilmos Hevesy, a Hungarian who was a technical manager in Bleriot's rapidly expanding airplane factories. Figure 2-17 shows the public announcement of Bleriot's attempt.

With his Budapest flight, Bleriot beat out a Hungarian ... Dr. Áugoston Kutassy ... as the first to fly a heavier-than-air craft in Hungary. After the Reims air meeting in France, Kutassy purchased a Farman-Voisin aircraft. At the time of Bleriot's Hungarian flight, Kutassy was taking flying lessons in France in preparation for a flight attempt in Hungary. In any case, Kutassy is recognized as the first Hungarian airplane pilot.

In addition to Kutassy, within a few months of Bleriot's first flight in Hungary János Adorjan,

Figure 2-17 "Bleriot in Budapest"

Mihály Székely, and Aladir Zsélyi also made successful flights within Hungary.

The Hungarian Aero Club was formed in 1902. In 1910, its members had developed enough confidence to conduct an *International Flying Competition* patterned after the famous 1909 event at Reims, France. It was held at Rákosmezö from June 5th to 15th, 1910. Because of their greater experience, the competitive events were dominated by the foreign pilots who participated.

The flying competition was immensely successful, with a very large attendance. A separate chapter on the *Flying Competition of 1910* provides additional information.

In these early years of aviation, prizes were frequently offered by the Austro–Hungarian government and commercial interests for "record" flights. On November 8, 1910, Capt. Petroczy won a 50km competition in Wiener–Neustadt. In December, an Austrian pilot won a prize of 4,000K for a cross–country flight from Wiener–Neustadt to Sopron and return. On April 18, 1911, Lt. Bier in an Etrich monoplane won a race over the same course in 48 minutes, during which he set an altitude record of 1,100 meters. On July 7, von Umlauff in a Lohner–Daimler biplane won the 15,000K "Times" prize for the first non–stop flight from Wiener–Neustadt to Budapest, in a time of 2 hours and 10 minutes.

Two 1911 flights (by Hungarian competitors in the 1910 flying competition) are purported to have carried newspapers and/or mail. No mail carried on these flights has ever been identified.

Mihály Székely, as a part of his first Budapest–Vienna flight, carried newspapers from Budapest to Györ -- on July 23–4. Yes, two days for the 70 mile flight to Györ! On the 23rd, shortly after takeoff he made an emergency landing at Solymár for engine repairs. He took off and made another forced landing at Esztergom for further repairs. From there he flew to Komárom for an overnight stay. On the 24th he flew to Györ and delivered his newspapers. On the 25th, he flew from Györ over Magyaróvár and Pozsony before landing at Fischamend. Finally, that same day he completed the last leg from Fischamend to Wiener–Neustadt.

Also in July, Antal Lányi planned to fly with newspapers and mail from Debrecen to Nagyvárad. After flying about 50km, he made a forced landing at Szalárd. From there his cargo was returned to Debrecen by road. In all probability, it was placed into normal postal channels. Certainly, none is identifiable.

In 1912, Guido Prodam attempted a flight from Fiume to Lovrana, across the gulf now known as Kvarner. He is believed to have carried a small number of postcards bearing his picture and signature. He crashed into the Adriatic. Fishermen rescued him and his postcards were mailed from Abbazia. The unstamped cards were delivered with postage due. Only one example is known to have survived.

There also exists flown mail from various airplane flights in other countries, particularly Germany, which was addressed to Hungary. Such mail, like the Zeppelin mail mentioned above, was forwarded to Hungary by surface means on the completion of the flight. This foreign airmail, which was ultimately delivered in Hungary, makes an interesting addition to a Hungarian airmail collection.

The state of the development of military aviation in Austria–Hungary was far behind that in France, Germany, Britain, and the United States. In June 1912, a central aeronautical committee under Prince von Fürstenberg was formed to create a national air fleet. One of its first recommendations was the purchase of six Zeppelins to protect the major cities of the Empire, including only Budapest in Hungary. The first of these, the 22,000 cubic meter M–IV was soon under construction, but the war came before its delivery.

The procurement of aircraft was so poorly financed by the Austro–Hungarian government that tea parties were held by prominent Viennese women to raise money to buy airplanes. Postcards were also printed and sold

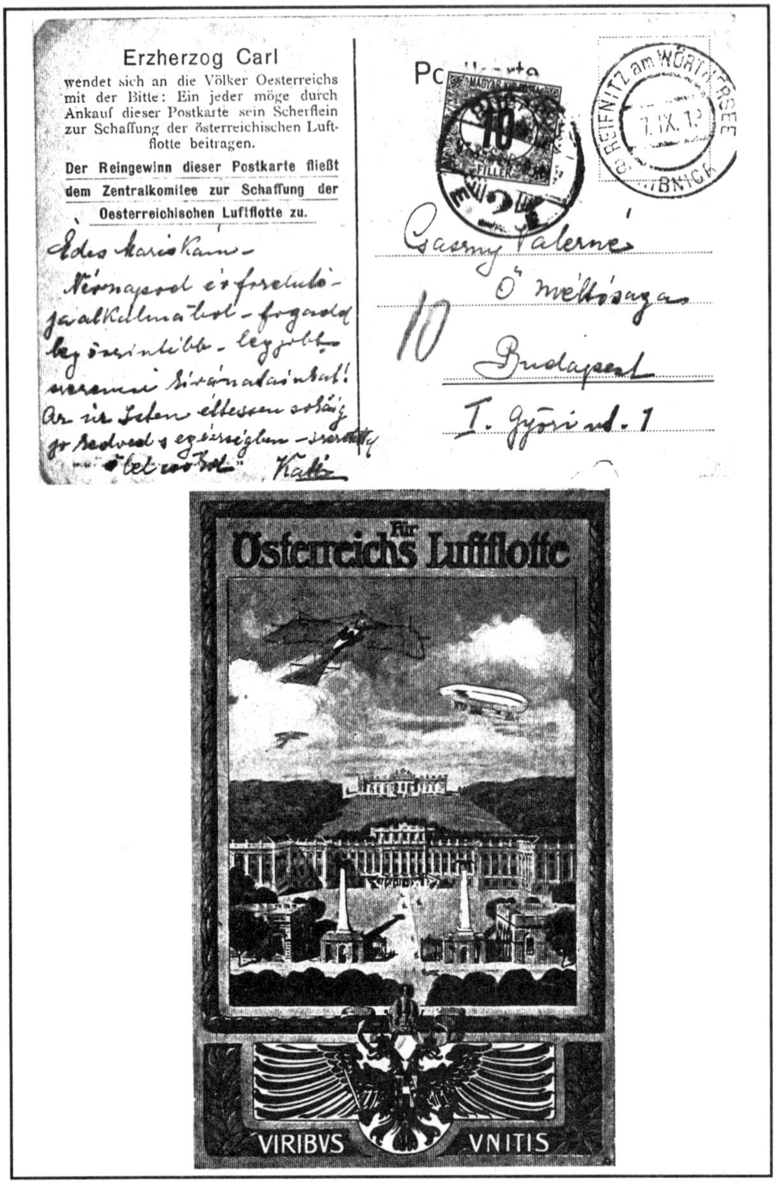

Figure 2–18 Card Appealing for Air Fleet Funding

to raise funds or request donations. Among them was the card shown in Figure 2–18. It makes a request for support on behalf of Archduke Carl, who was to become Hungary's last king. Note that the card was mailed in Austria without postage, so Hungarian postage due was assessed.

At the end of 1913 there were only nineteen qualified pilots in the military. The army owned 26 planes, and the navy four.

In April of 1914, a long cross–country flying competition was held for the *Schicht Prize* of 100,000K. Army and civilian pilots participated in this circular course through both Austria and Hungary. A postcard in the Heller correspondence is likely to be a report from the von Pischoff plane in the *Schicht Prize* competition. A red and black vignette is the only philatelic commemoration of the competition.

By the end of 1914, the number of army planes had increased to 110. They were largely of German design, manufactured by the Austrian branches of large German firms. The navy's fleet of seaplanes had grown substantially, and the naval pilots were renowned for their flight operations. This was largely due to the quality of the naval aviation school at Cosada. By this time, of course, Europe was at war.

Army Balloon Mail of 1898–1914

The first of the army aeronauts to carry mail aboard his balloon flights was Gustav von Tepser. Picture postcards were printed showing ten different military ballooning scenes. During the summer of 1898, Tepser carried

Balloon Cards Used by Tepser

No.	Description
1	Füllung – inflating
2	Auf dem Manöverfeld – in the field
3	Umfüllung – transfering gas
4	Vor der Abfahrt – before the ascension
5	Los! – release!
6	In den Lüften – in the air
7	Hoch transport – travel on high
8	Einholen – hauling in
9	Glüchliche Reise – have a good trip
10	Nach der Landung – after the landing

a few of these cards aboard each of his flights. He posted the cards where he landed. Most were posted in Hungary and bear Hungarian stamps. The example shown in Figure 2–19 is number 4 of the series.

Tepser's manuscript notations on the cards are difficult to read. They are written in the Gabelsberger shorthand system used at the turn of the century in southern Germany, Austria, and Hungary. The flown Tepser card shown in Figure 2–19 was addressed to his sweetheart Bertha Palatka. He posted it on the Boba–Csaktornya railroad. Tepser's notation reads: "Landed safely at Andráshida near Zalaegerszeg. A thousand kisses as always, your devoted Gustav." The flown Tepser cards are extremely scarce.

Figure 2–19 Tepser Card of 1898

Tepser's cards fall into one of the three categories of balloon–related Hungarian mail of the period. This type is usually called "landing greetings" mail. The three categories are:

Landing Greetings Mail. This mail was posted where the balloon landed; usually bears a message concerning the flight; is often addressed to other aeronauts or to a wife/girlfriend; and *may* have been flown aboard the balloon. In Tepser's case, the cards were obviously flown prior to posting.

Army Maneuvers Mail. This mail was posted from army maneuvers in the field; may picture balloons or other aeronautical motifs; may have been officially prepared by Army authorities as greeting cards from the maneuvers; may have been from members of balloon units; but, almost certainly was *not* flown. Figure 2–20 is an example from the 1901 maneuvers. It bears a fieldpost cancellation *KuK Feldpostleitung No. 13.* The message reads, in part, "We are participating in the maneuvers and getting considerable air time."

Figure 2–20 Maneuvers Card of 1901

Figure 2–21 shows a 1911 card from the maneuvers in Sztropkó in northern Hungary. Its reverse pictures various scenes -- including the three modes of military aviation: balloon, dirigible, and airplane. It also shows the senior military leaders of Austria–Hungary including Army Chief of Staff Ferenc Conrad, Count of Sonnenstein and Hötzendorf who happens to be

Figure 2–21 Maneuvers Card of 1911

Czinfalva. Therefore, postage due was assessed.

In 1908, officially prepared balloon postcards went into use as "drop cards." The front of the cards is rather standard except that the word "postcard" is printed in ten languages. The reverse of these cards was printed with the heading *Glück Ab!* or *Glück Ab! Gut Land!*, a small picture of a balloon, and what was in essence a form to be completed by the pilot and the finder.

a relative of this author. It would be unusual to find a *maneuvers card* that was flown.

Balloon Drop Mail. This mail was carried by balloonists and dropped in flight. It was presumably posted by the finder and usually has a notation from the pilot. It sometimes also has a notation by the finder. This mail was obviously flown.

Virtually all of this balloon-related mail consisted of postcards franked by stamps. The balloon drop cards are of most interest to airmail collectors, but the other mail mentioned also represents a significant aspect of the aerophilatelic history of Hungary and some of it was flown.

Drop mail posted before 1908 is found on various picture postcards, although the reverse often shows an aeronautical or ballooning view. This mail is quite scarce. The 1903 drop card shown in Figure 2–22 has a photo of the army balloon station at Steinfeld on the reverse. It was franked with an Austrian stamp, but was found and mailed at the Hungarian border village of

The colloquial meaning of the headings was "Happy Landings!" The color of the cards was usually a pale blue or pale green. These cards

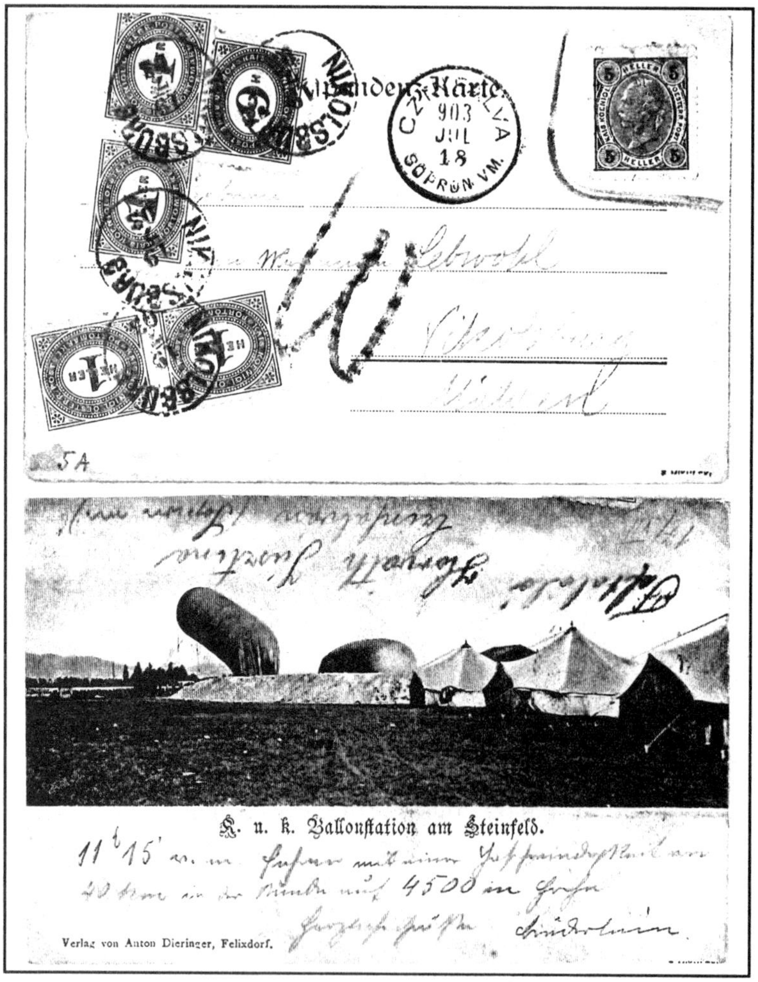

Figure 2–22 Balloon Drop Card of 1903

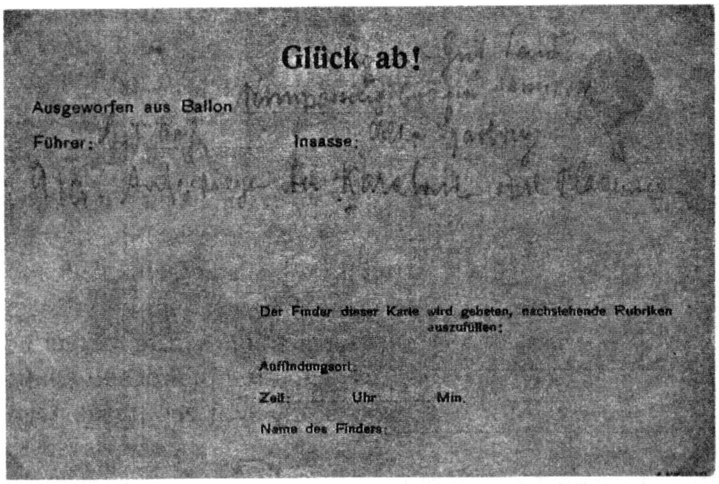

Figure 2–23 "Glück Ab!" Card from the *Wimpassing*

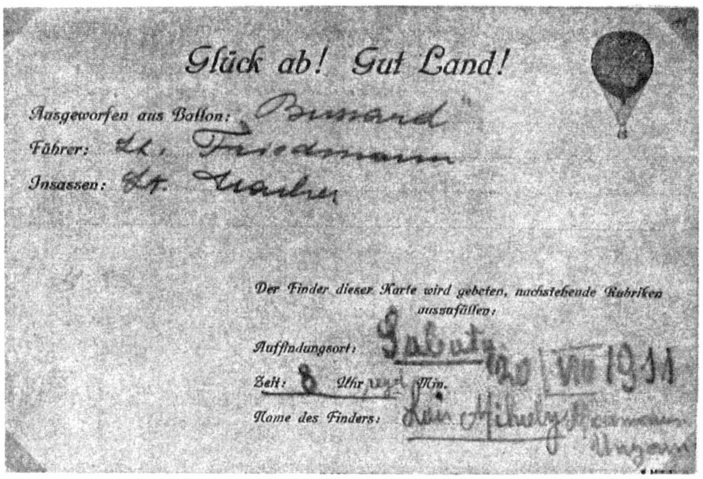

Figure 2–24 "Glück Ab! Gut Land!" Card of the *Bussard*

were used primarily from 1908 to 1911, but are occasionally found from as late as 1914. Figure 2–23 shows the "Glück Ab!" type dropped from the balloon *Wimpassing*. Figure 2–24 shows the "Glück Ab! Gut Land!" type dropped from the balloon *Bussard*.

Philatelic authors who claim to exhaustively list these balloon flights have made a serious error. They have given the unfortunate impression that the cards can and should be collected by flight. There were only a few balloons, hundreds of flights, and just a handful of cards from each flight (if any). So, to attempt to list ... no less collect ... all the flights is impossible. This is evidenced by the fact that most such "catalog" listings consist of only six or eight balloon flights. Collecting by card type, or by the specific balloon, is a much more reasonable approach, although the scarcity of these cards makes even that extremely difficult.

This author has seen, or found reliable references, concerning balloons with the following names, with mail posted in Hungary: **Budapest** (also **Bp II**), **Bussard**, **Cattaro**, **Erzh. Josef Ferdinand**, **Excelsior**, **Hergesell**, **Hungaria** (also **II** and **III**), **Kibitz**, **Radetzky**, **Ragusa**, **Sirius**, **Uhu**, **Waag**, **Wien** (also **II** and **III**), and **Wimpassing**. Others very likely exist.

International Flying Competition of 1910

In June 1910, the Hungarian Aero Club sponsored a flying competition patterned after the famous 1909 event at Reims, France. The *International Flying Competition* was held at Budapest's flying grounds at Rákosmezö from June 5th to 15th.

During the competition, a post office was established at the flying grounds. A large circular commemorative cancellation was applied to mail there. The cancellation reads "Budapest Repülö–versenyek" (in translation, "Budapest Flying Competitions"). Mail with this cancellation was not flown. Nevertheless, such mail is among the most sought after in early Hungarian aerophilately. Figure 2–25 shows an official envelope of the competition with the commemorative cancellation.

The competition was plagued by bad weather. This was particularly a problem for the inexperienced Hungarian pilots. For their own safety, they were only allowed to briefly "hop" their planes off the ground to qualify for the

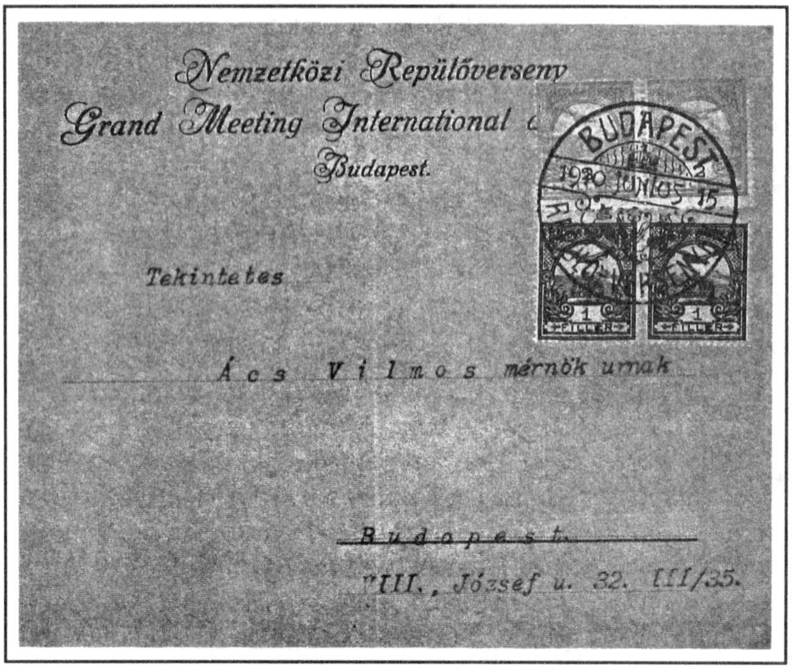

Figure 2–25 Flying Competition Cover of 1910

National Prize. As a result, Ernö Horváth won with a ridiculously short 8 second flight of 112 yards, followed by Székely and Adorján with flight times of 6.8 and 5.0 seconds respectively. In any case, not much was expected of the Hungarians against their experienced foreign competitors.

It was not only the Hungarian pilots who were plagued by the weather. The postcard to France shown in Figure 2–26 displays evidence of pervasive weather-related problems. It pictures a Farman biplane that had been "flipped" by the wind.

Ernö Horváth, the *National Prize* winner, was a young mathematics and physics teacher. He is remembered in aviation circles for the introduction of a universal steering control and a canvas covered fuselage on his monoplanes.

The 1967 airmail stamp in Figure 2–27 pictures Horváth's 1911 airplane with the enclosed fuselage. In the 1910 competition, he flew a Bleriot-type monoplane that he constructed himself, powered by a 24 hp Anzani engine. It is shown in Figure 2–28.

Figure 2–27 Horváth Stamp of 1967

To give readers some idea of the aviation capabilities of the day, here are some of the first place winners at the Budapest competition.

Duration: 2h3m47s by Wagner. **Distance**: 137.4 km by Wagner. **Speed**: 76.8 km/hr by Latham. **Altitude**: 1060 meters by Paulhan. **Shortest takeoff**: 11.05 meters by Paulhan (a world record at the time).

Figure 2–26 Flying Competition Postcard

One prize was not awarded. That was the *Cross Country Prize* offered to anyone who could fly from Budapest to Györ, a distance of about 100 km. The rules allowed only one stop, and imposed a 5½ hour limit. Only one competitor, the Austrian von Pischoff, attempted this flight. He gave up after his second forced

Figure 2–28 Horváth's Competition Airplane

Figure 2–29 Adorján's Competition Airplane

Figure 2–30 Kutassy's Competition Airplane

landing due to weather, and had his plane towed back to Budapest.

The airplanes of three of the Hungarian competitors are shown in Figures 2–28 (Horváth), 2–29 (Adorján), and 2–30 (Kutassy).

The actual flying competition took place only through June 10. The air show continued until the 15th with a variety of demonstrations. The climax was the simultaneous ascension of three Army balloons from the flying grounds. Figure 2–31 shows a photo of that ascension by balloonists Nemec, Hirsch, and Stohanzl, before a crowded grandstand.

The balloonists reported to headquarters in Vienna that they flew for "2 hours and 30 minutes to altitudes of 1100 meters." The "landing greetings" card bearing that report, on an official postcard of the competition, was signed by all three balloonists. Figure 2–32 shows the reverse of this card.

A vignette, printed in four colors, was issued to promote attendance at the competition. Its design was very similar to the card shown in Figure 2–32. These vignettes are normally not found on mail with the competition postmark.

It should also be noted that a set of postcards picturing events of the 1910 flying competition were produced at a later date. These are often found with the cancellation of the *1917 Army Airplane Exhibit*. These cards should not be confused with the mail posted at the flying grounds in 1910.

Figure 2–31 Triple Balloon Ascension

Figure 2–32 Balloon Ascension Report

To give an idea of the range of this material that is available, two cards are illustrated. The 1902 card shown in Figure 2–33 is addressed to my grandmother, Eszther Nemes. It shows "Airships of the Year 2000." This card, and knowing Igor Sikorsky personally, were the initial motivations for my interest in pioneer aviation. Other than the personal nature of the transportation depicted in this 1902 card, aviation progressed much faster than conceived by the artist.

A Touch of Fantasy

One often neglected facet of aerophilately are the "fantasy" cards that were so popular in the early part of the century. These were postcards that pictured how people expected aviation to progress during the 20th century. They were certainly not flown, but nevertheless add an interesting perspective to the philatelic history of aviation. These cards range from true science fiction, to photo shop productions of children "flying" in airplanes, to predictions that actually took place much sooner than anticipated.

Figure 2–34 Aviation Fantasy Card of 1914

Many of the aeronautical fantasy cards of the early 1900's were advertising cards. The well–known Budapest printing firm of Emil Winter was among those who produced aviation fantasy cards for advertising purposes.

The 1914 card shown in figure 2–34 has a different purpose ... war propaganda. It is titled "On the Towrope" and shows a Zeppelin easily toppling the Eiffel tower while ineffective ground fire falls short. Its picture of a quick, easy war was not to be –– as we shall see in the next section.

Figure 2–33 Aviation Fantasy Card of 1902

Aviation and Airmail of World War I

The *Great War* lasted over four years. It began with the assassination of Austria–Hungary's heir–apparent and ended with the dissolution of the Dual Monarchy. World War I is generally viewed as a German–British–French (and later an American) affair. Many historians treat the eastern and southern fronts as a backwater. Yet this is where hundreds of thousands of Hungarians fought and died.

Before the war began, as previously noted, the military aviation capability of Austria–Hungary was not up–to–par for a major power. Britain, France, the U.S., Germany, and even Russia had far superior air forces. In von Kármán's words, the Austro–Hungarian Air Force was a "joke." With the coming of the war, the Austro–Hungarian air corps built up very rapidly and some significant innovations were made. But, the worldwide development of military aviation was on an even faster track. In most aspects of aviation, Austria–Hungary was not a leader. Airmail may have been the exception.

A big question for many Hungarian aerophilatelists concerning World War I military material is "What should I collect?" My answer is an unequivocal "Collect what interests you!" I chose a rather broad standard in my collecting and exhibiting. I included unflown material related to military aviation ... though in exhibiting I played these items down somewhat. A standard to which I adhered was the "Hungarian connection." I required that each piece must meet at least one of three criteria: (a) it originated in Hungary; (b) it was destined for Hungary; or (c) it displayed a Hungarian language postal marking. In fact, most of the WWI mail in my collection meets two of these three criteria. The Hungarian word for "postcard" or "sender" being printed on the card is not sufficient for me. Austria–Hungary was a multi–lingual state. Again, collect what interests you!

I have chosen not to provide many details on the aircraft that were used by the Austro–Hungarian forces. That data is widely available in the specialist literature.

Aviation During World War I

The Austro–Hungarian army and navy air forces were very small at the war's beginning; but, they grew rapidly.

The basic unit in the Army's *Fliegertruppen* (Air Corps) was the *Fliegerkompagnie* ... literally "Aviator Company." These were self–contained units consisting nominally of six airplanes, 15 officers, and 117 NCO's and enlisted men. They were often assigned to an army headquarters as a *Fliegerabteilung* (Aviator Detachment). Figure 3–1 shows an army biplane at a field in Serbia early in the war.

An *Etappenflugzeugpark* (Reserve Airplane Base) was an air corps reserve base. It nominally maintained three airplanes, two pilots, and an observer for each Aviator Detachment in its area of responsibility. The *Fliegerersatzabteilung* was the aviation replacement unit.

Another duty of the army air corps was the operation of balloons. The primary role of most balloon units was artillery spotting. A few on the Adriatic were involved in harbor defense. After the first couple of months of the war, Austria–Hungary did not operate any military dirigibles. The balloon units were typically called *Luftschifferabteilung* (Airship Pilot Detachment) or *Ballonkompagnie* (Balloon Company). The air corps was also responsible for the operation of each *Feldwetterstation* (Weather Station).

Figure 3–1 Army Biplane in Serbia

A normal fieldpost office operating in the vicinity usually processed the mail of an army air corps unit. Figure 3–2 shows a typical fieldpost "office."

The personal and official mail of members of the air corps units was handstamped with a unit postmark. Most of this mail was *not* flown to its destination. The exception is some official mail that was forwarded to headquarters by courier flights. On occasion, one can positively identify these flown letters by the date and time of their postmarks and receiving marks. The chapter on *Army Aviation and Balloon Units* discusses the postal markings of the various air corps units.

The navy air wing had responsibility for rapid communications along the rugged Adriatic coast. It also took offensive and defensive actions against enemy fleets, and offensive action against Italian ports on the north and west shores of the Adriatic.

A few small planes were used for courier services from land bases, but flying boats were the mainstay of the Austro–Hungarian Navy's *Seefliegerkorps* (Seaplane Corps).

The headquarters of the Seaplane Corps was at naval headquarters in Pola, near the southern tip of the Istrian peninsula. Here, in addition to headquarters, there was a seaplane station at St. Catarina island within Pola harbor itself, and at Puntisella at the harbor entrance. The Navy's *Schulflugstation* (Flying School Station) was on Cosada, a small island just outside Pola harbor.

Figure 3–2 Army Fieldpost Office

Another major seaplane base operated at Kumbor on the Gulf of Cattaro. Here the mission was largely defensive ... to spot submarines attempting to sneak into this large Austro–Hungarian and German naval base. Upward of a dozen other smaller seaplane facilities operated at one time or another on the northern and eastern shores of the Adriatic. They ranged from as far north as Grado -- just a few miles from Venice, Italy -- to as far south as Durazzo, Albania. At different times during the war, these were called *Seeflugstation* (Seaplane Station), *Seeflugstützpunkt* (Seaplane Base), or *Küstenflugstation* (Coastal Air Station). The seaplane station at Fiume was the only one of these facilities on Hungarian soil. Figure 3–3 shows a typical flying boat of the Austro–Hungarian Seaplane Corps.

The navy operated two other seaplane facilities. A seaplane test station was located at Keszthely on Lake Balaton. Also, during the last few months of the war, the navy operated a seaplane base at Odessa in occupied Ukraine. No aerophilatelic material is known from either of these bases.

Most of the naval seaplane units on the Adriatic had their own unit markings. Both personal and official mail was handstamped with this marking. The naval postoffice at Pola handled all of the mail of the Seaplane Corps personnel in the immediate Pola vicinity. It was rarely, if ever, flown. There were no special naval mail facilities at the other seaplane bases and stations. A given piece of mail might have been handled by the civilian postoffice, forwarded by ship, or flown either by a courier flight or a routine mission to Pola or Kumbor. Mail with markings that indicate it was forwarded by a courier flight is extremely rare.

Due to the remoteness of many seaplane bases, it was not unusual for a pilot to carry mail for men in their unit. Therefore, unlike the Army, a small but significant percentage of the Navy Seaplane Base mail was flown. Positive identification of flown mail is only rarely possible. The best indicator that an item was flown is that it originated at one of the smaller bases *and* it shows the postmark of either Pola (*Marinefeldpostamt*) or Kumbor (the civilian postoffices at *Teodo* or *Zelinika*).

Most of the aircraft used by the Austro–Hungarian forces were manufactured in Austria by subsidiaries or licensees of German aircraft companies. There were only three airplane manufacturers in Hungary.

The Hungarian Lloyd Airplane Factory (*Magyar Lloyd Repülögép Gyár*) of Budapest manufactured Lloyd biplanes used by the Army. Pioneer pilot Henrik Bier managed this factory.

Figure 3–3 Typical Seaplane Corps Flying Boat

The Hungarian Airplane Works (*Ungarische Flugzeugwerke AG* ... known as UFAG) was located in the Albertfalva section of Budapest. The company was formed in 1912 by a merger. Ernst Heinkel was the chief designer. In 1915 they began building Lohner flying boats under license. These were tested on Lake Balaton, at the navy's Keszthely seaplane station, before being sent into action on the Adriatic.

The Hungarian General Machine Company (*Magyar Általános Gépgyár RT* ... known as MAG) built Fokker aircraft and engines for the Army near Mátyásföld, Budapest's airport.

One of the seven chapters that follow covers virtually all of the mail of the WWI period that is of aerophilatelic interest. I will mention a couple of exceptions here.

One can find occasional mail posted in Hungary, with German air force unit markings (other than of the *Zeppelin Base near Temesvár* that has its own chapter). I have seen such mail posted at Budapest, Szolnok, and Herkulesfürdö (now Baile Herculane, Romania). These cities may all have been part of a German air courier route from Germany to Turkey. The Hungarian airmail collector is likely to treat this German mail in the same manner as unflown mail with aviation unit markings originating in Hungary.

Though not actually mail, other flown missives include the propaganda leaflets dropped by allied forces late in the war. Those used on the southern (ie. Italian) front included messages in Hungarian. Some aerophilatelists find these of peripheral interest.

Przemysl Airmail of 1914–15

The Siege. Przemysl was an important fortified city on the river San in Austrian Galicia. The Russian advances into Austria-Hungary early in the war immediately put it at risk. Przemysl was defended by about 127,000 troops, the majority of them Hungarian. Przemysl was a *modern* ring fortress, with a 30-mile perimeter holding 19 entrenched fortresses and 23 smaller forts, batteries, and fieldworks.

It was also a rail and communications center. Many considered it the key to the control of Galicia. The fortified area of about 60 square kilometers, consisted of the city itself and numerous nearby villages. In all, about 20,000 civilians lived within the fortified area.

The two highest ranking officers at Przemysl were both Hungarians. General of the Infantry Hermann von Kuzmanek was in overall command. He was born in Nagyszében in 1860 and died in Vienna in 1934. Figure 3–4 shows Kuzmanek in 1914 with his handlebar mustache in its full glory.

Figure 3–4 Gen. Kuzmanek

In command of the Hungarian National Guard forces (the "Honvéd") was Lieutenant General Arpád Tamásy. He was born at Köhalom in 1861. Figure 3–5 pictures the Przemysl command staff with Tamásy in the inset.

Figure 3–5 Przemysl Command Staff

Przemysl first came under siege by the Russians on September 18, 1914. Immediately, communication was a concern. There was only a single radio in the city and at that time radio communication was notoriously unreliable and subject to intercept. At times they were able to flash signals to the Austro-Hungarian Third Army in the Carpathians, forty miles away. Also, an aerial courier service was quickly put into place using *Fliegerkompagnie No. 11*. The first siege was relieved on October 10. Only one flight into and out of the city took place during the first siege. On November 7, the Russians besieged Przemysl for the second and final time.

Between the sieges, it took too long for the Austro-Hungarians to get organized and repair the railroad between the Carpathians and Przemysl. As a result, it was only during the week of October 28 to November 4 that the city could be resupplied. In all, 128 trainloads of food, ammunition, and coal arrived. The Austro-Hungarian advances were repelled. Army Chief of Staff Conrad ordered a general retreat and Przemysl was again surrounded.

During the early weeks of the second siege, the Austro-Hungarians had the advantage of much superior artillery. They easily kept the Russians at bay. Time, though, was on the side of the Russians.

On December 15th, Gen. Tamásy attempted a breakout that coincided with the descent of the Third Army from the Carpathians. With 17 battalions he broke through the investment lines, but within two days ... and before they were able to link up ... the Russians defeated both him and General Krautwald of the Third Army. Tamásy retreated to Przemysl.

In one of the last examples of the chivalry attributed to "gentlemen at war," a truce was called on Christmas day. The Austro-Hungarian officers invited their Russian counterparts to the edge of the city and offered them drinks and refreshments. A military band entertained. The respective commandants eulogized the bravery of the opposing armies. They drank toasts to both Tsar Nicolas and Emperor Francis Joseph. The next day, the war went on.

Soon after the holiday it became apparent that the food supply of Przemysl was rapidly coming to an end. Horses were slaughtered for meat. This further damaged the mobility of the army that was essentially without motorized vehicles. In late January, an attempt to relieve Przemysl by the combined German and Austro-Hungarian armies was again repulsed -- this time more by the winter weather than the Russians. The Third Army suffered 68% casualties. Many froze to death.

Another attempt at relief failed in early March and by March 15 Fieldmarshal Conrad recognized the reality that Przemysl would fall. With firm information that all food in the city would run out on March 24, Kuzmanek made one last desperate attempt at a breakout of the entire garrison.

Led by Tamásy's *23rd Honvéd Infantry Division*, the attempt was a miserable failure resulting from both bad weather and the wretched state of the troops due to hunger and scurvy. Mass destruction of the fortress' military assets began on the 19th and the last artillery ammunition was fired on the night of the 20-21st. The cannon barrels were then blown up. On the morning of the 22nd, white flags were planted around the city. The entrenched fortresses themselves were blown up.

Figure 3-6 POWs at Przemysl

A deputation was sent out to formally surrender to the Russians later that day. In all, the Russians took into captivity nine generals, 93 staff officers, 2,500 officers, and 117,000 men. For most of the next two weeks, 10,000 men per day were marched off to prisoner of war camps in Russian Turkestan and Siberia. Figure 3-6 shows part of that march.

Less than three months later, on June 2, Przemysl was recaptured by combined German and Austro-Hungarian forces. It was held until the war's end. Many of the prisoners of war were unable to return to Hungary until the early 1920's.

Figure 3-8 Przemysl Forerunner Card -- Type B Marking -- Flown to Kassa on December 9

Forerunner Airmail Service. As mentioned above, a single airplane made it to Przemysl during the first siege. Capt. Taussig piloted the plane, with Capt. Raabl as an observer. In addition to military dispatches, as a favor to besieged officers, the pilot carried some personal mail out of Przemysl when he left on October 6. He set the precedent for the "Przemysl Forerunner Airmail Service" that flew additional mail out as favors until the official airmail was instituted in January 1915.

That first flight made a forced landing at Jaslo. Taussig was injured in the crash. Raabl brought the mail to headquarters by auto. There it was canceled on October 7 by *Feldpostamt 11*. There were about 100 pieces of such mail in all. I have never seen nor heard of mail from this flight that was addressed to Hungary. Some certainly may exist.

During the 1914 phase of the second siege, flights out of Przemysl carried "forerunner" airmail on three separate dates, as favors to officers. This mail was mostly regular Fieldpost postcards. On November 23 (often referred to as Flight 2), Capt. Lehmann flew mail from Przemysl to Cracow. On December 9 (Flight 3), Fieldpilot Wanneck flew from Przemysl to Kassa ... the only flight that landed on Hungarian soil. On December 26, two planes piloted by Lts. Schmocser and Rzemenowski flew to Cracow (Flights 4 and 5).

One can identify mail from these flights by the dates of the receiving marks and by the unit markings of *Fliegerkompagnie 11*, which flew the mail. There are three unit markings of particular interest from these flights. The first, which I call type "A" is an 85 mm long straight-line handstamp that reads *K.u.k.*

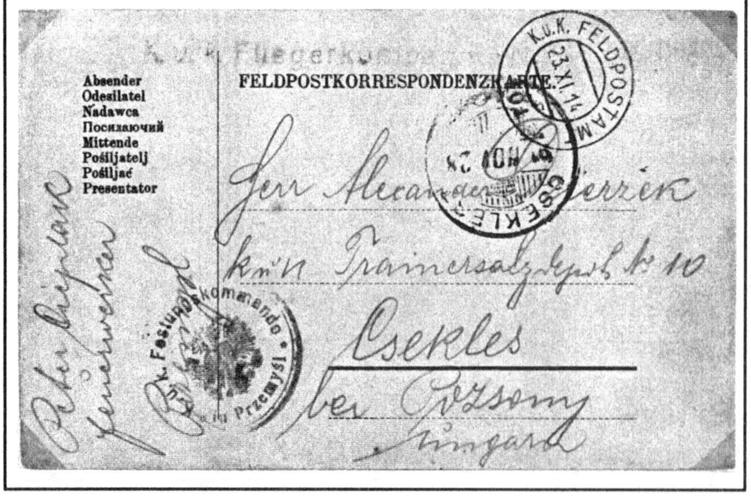

Figure 3-7 Przemysl Forerunner Card -- Type A Marking -- Flown November 23 to Cracow

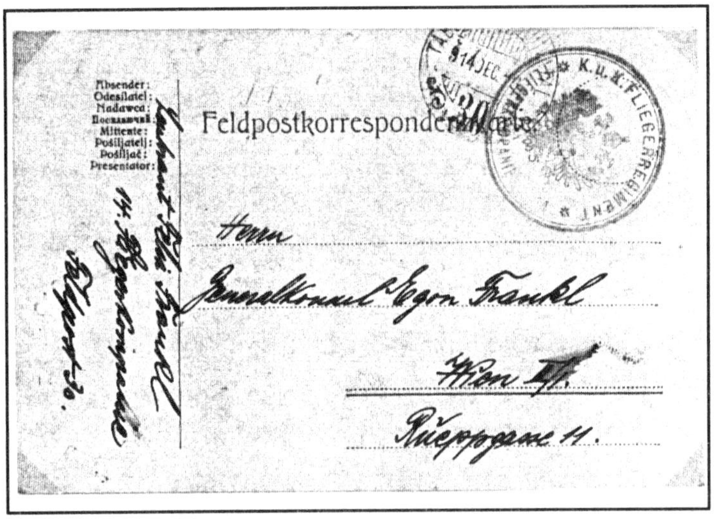

Figure 3-9 Card showing Przemysl Type C Marking

Fliegerkompagnie Nr. 11. It is usually lilac, and often faint. This marking can be seen in figures 3–7 and 3–10.

The second marking is a circular handstamp of the same unit that I call type "B." It is 31 mm in diameter, has the Habsburg eagle in the center, and also reads *K.u.k. Fliegerkompagnie Nr. 11* around the periphery, inside the circle. This marking was also applied in lilac. It is shown in Figure 3–8.

One *or* the other of these markings ... Type A or B ... *must* be on every forerunner airmail card flown out of Przemysl during the second siege. The cards also show the unit marking of the sender's unit. Examples of such unit markings are shown in Figures 3–7, 3–8, and 3–10.

The mail of the November 23rd flight entered the normal fieldpost system in Cracow at FPO 101 on the same day. Many of the pieces bear the *K.u.k. Feldpostamt 101* postmark as well, for example the card shown in Figure 3–7. Delivery of this mail at its destination began on November 27.

The mail flown on December 9th was passed to FPO 30 in Kassa that same day. All of the mail from this flight has the *Tábori Postahivatal 30* postmark of the 9th. Unfortunately, this cancellation was not carefully applied. Therefore, it is often weak or partial. Figure 3–8 shows an example. The mail of this flight is the most sought after Przemysl airmail by Hungarian collectors.

A few of the cards carried on the flight of the 9th were subsequently handstamped with the forerunner mark that I call Type "C." It is a 35 mm circular unit marking of *K.u.k Fliegerregiment / Fliegerkompanie No. 14*. This postmark was applied in Kassa. The Przemysl airmail destined for Austria was flown to Vienna, on the 10th, by an Aviator Company 14 pilot.

Since this marking was applied only to cards destined for Austria, it is not seen on Przemysl airmail addressed to Hungary. Figure 3–9 shows the Type C marking on a postcard that did not come from Przemysl. It originated in Kassa on December 8, and may also have been carried on that same courier flight. I expect that being an officer, a pilot, and related to a general helps you get better service for your personal mail!

Which of the two airplanes, that made the trip out of Przemysl on December 26, carried the

Figure 3-10 Przemysl Forerunner Card -- Flown with First Regular Airmail

mail is not known. The mail was turned over to FPO 186 on December 27, and some of it was postmarked by them. It was delivered in Vienna on December 30 and to Hungary the following week.

Shortly after the new year of 1915 began, the arrangements were announced for a regular airmail service to carry mail from Przemysl. A small number of forerunner–type cards were prepared prior to that announcement. These were carried on the first regular flight of January 18, but bear the markings of the 1914 forerunner mail. They were carried by the pilots as "favors" along with the "regular" airmail carried that day. The example shown in Figure 3–10 is handstamped with the Type A marking. It was delivered in Hungary on January 22.

It is important to remember that Aviator Companies 11 and 14 continued to operate throughout the war, and used the same unit markings. Certainly every piece of mail with their markings did *not* originate in Przemysl, as some have imagined. The number of pieces carried on each flight of the Przemysl forerunner airmail is small. This mail is extremely scarce and quite desirable.

Regular Airmail Service. The Przemysl regular airmail service came into being on January 8, 1915, when the base newspaper -- the **Tábori Újság** -- announced the regulations under which it would operate. It ended with the surrender of Przemysl on March 22. While the word "regular" is applied due to the official regulations, from a schedule perspective it was hardly regular. Actually, it was quite sporadic. This results from both the military situation and the other higher–priority tasks assigned to Aviator Company 11 ... including bombing runs against the Russians. Here we only focus on the mail–carrying flights, not on any of the other missions. Figure 3–11 shows an *Albatros* biplane of *Fliegerkompagnie Nr. 11* at Prezemysl preparing for takeoff with the mail.

The Cards. All regular Przemysl airmail used official postcards that were sold for 50 fillér. The proceeds went to a trust fund for war widows and orphans. While officers could purchase cards individually, only one card was allocated for every five enlisted men. It was recommended that men from the same towns send cards jointly. Few complied with the rule concerning five men per card, but some cards with 5, 10, or even more short messages are known.

The cards were numbered for control purposes. No military matters could be included in the messages, and no codes were permitted. The cards were censored twice, and on the first of these occasions they were signed by an officer.

No unit identification could be included in the message. Instead of this, a *Fliegerpost Przemysl* handstamp was applied. Also, no regular fieldpost cancellation was used. Rather, the *IX/54* circular postoffice designator of Przemysl was applied. The intent of these rules was to minimize the risk, in the event of mail being captured, of the Russians identifying the precise units that were within Przemysl. The rule about unit markings was occasionally violated, most frequently by the headquarters command staff and the balloon detachment. The other rules were carefully observed.

Figure 3–11 Aviator Company 11 Biplane at Przemysl

Of the four markings that are found on every Przemysl airmail card, three of them: the numbering, the *Flieger–*

post Przemysl marking, and the *IX/54* marking were applied prior to sale. The censorship markings were obviously applied later. As a result of this policy, unused cards with most of the markings exist -- probably carried to Russia by prisoners. These unused cards are an obvious potential base for "created rarities." So, exercise care.

The postcards used were made of very thin paper, and as a result many have not worn very well. This type of card is generally referred to as a *flimsy* -- and it is! On the face of each, in addition to *Feldpostkorrespondenzkarte*, the word for "sender" is printed in the eight languages of the Empire. Tranmer estimates that about 90,000 cards were printed. The percentages used as intended, carried into captivity by soldiers, or destroyed prior to the surrender is unknown. The cards were initially produced in three colors (yellow, green, and rose) using a very soft, light–weight paper. Toward the end, an even lighter–weight ... but hard and stiff ... grey–green paper was also used. On rare occasions the blue cards prepared for the balloon mail were used for the airplane airmail. Tranmer notes that an experimental two–ply paper, with the front colored a deep rose and the text side white, was briefly tried.

According to Clement and Sobetzky, the color breakdown of the cards flown by the regular Przemysl airmail was as follows: yellow 38%; green 26%; rose 24%; gray–green 11%; and blue 1%.

Please note that the normal rose–colored cards are quite sensitive to sunlight and that the color has often faded over the years. Often they now appear to be buff to light brown.

The Regular Airmail Flights. The first occasion when regular airmail was flown out of Przemysl was on January 18. Three planes left that day for Brzesko. These flights, piloted by Lts. Schindler, Schmocser, and Stanger, are often referred to as airmail Flights 6, 7, and 8. They carried a large number of the new lightweight cards, plus some forerunner mail as favors. Mail carried on this flight is dated from December 25 to January 16. It is not known which plane(s) carried the mail. A small percentage of this mail bears the double–ring transit mark of *Moravska Ostrava* of January 20.

On February 7 (Flight 9), Capt. Lehmann flew the mail from Przemysl to Cracow. This mail was dated between January 17 and February 5. Some of it was postmarked in Cracow by FPO 56 or FPO 101.

On February 10 (Flight 10), Lt. Schindler flew the mail to Brzesko. This mail was dated February 6 and 7. Tranmer estimates that about 25% of it was postmarked by FPO 56 on the 11th.

On February 12 (Flight 11), Capt. Lehmann flew mail dated between February 8 and 10 to Brzesko. About half the mail from this flight was postmarked at *Moravska Ostrava* on the 13th.

On February 18 (Flight 12), Sgt. Meltsch again flew the mail to Brzesko. This mail was dated between February 11 and 16. It was again passed to the postoffice at Moravska Ostrava, where some of it was canceled on February 20 and 21.

On March 22, the day of the surrender, Lt. Stanger piloted a plane out of Przemysl with a large amount of mail, dated from February 17 to March 20. He landed at Brzesko, and some of the mail was postmarked by FPO 11 on the 23rd. This mail was considerably delayed before delivery in Austria and Hungary in May and June.

Also on March 22, Sgt. Meltsch left Przemysl with mail. He made a forced landing behind the Russian lines and almost all the mail he carried was lost. Tranmer reports that a few pieces surfaced, and bear Russian Red Cross markings.

Tranmer believes a backlog of unflown mail was captured by the Russians at the Przemysl postoffice. If this mail was somehow preserved and entered the philatelic market, it would be indistinguishable from flown mail.

Basic Przemysl Airmail Markings. The two most basic Przemysl regular airmail markings are the two–line *Fliegerpost Przemysl* handstamp that occurs in six types, and the circular *IX/54* postoffice designator that occurs in five types. Identifying the types of these two markings on virtually every Przemysl airmail card is easy to do.

Figure 3–12 -- which, to aid identification, was *not* reduced -- shows the **Type 1** *Fliegerpost Przemysl / Jänner 1915* marking in combination with each of the five circular handstamps. This 67 mm two–line marking is

Figure 3–12 Types of *IX/54* Circular Przemysl Marking -- Types a, b, c, d, e (top to bottom) -- all with *Fliegerpost Przemysl* Marking Type "1"

obvious because it is the only type that includes the word *Jänner*. The other five types of the *Fliegerpost Przemysl* markings are shown later.

The circular Przemysl postoffice designator is not unlike the many such handstamps used internally at postoffices throughout the Empire. The *IX* denotes the province of Galicia; the *54* is the number of the Przemysl postoffice. Figure 3–12 shows each of the five types of this marking. I designate these types by lower-case letters (a, b, c, d, and e) to eliminate the confusion caused by other authors who designate them with Roman numerals. This marking was usually applied in the upper right-hand corner of the card.

Type a of the circular marking is obvious to identify. The circle is only 20 mm in diameter. The diameter of all the other types is 26 mm. Types "a" and "d" are the commonest types of this marking.

Type b of the circular marking is 26 mm. in diameter. The circle is thin-lined (about ½ mm thick) and the *IX* measures 13–14 mm. wide.

Type c of the circular marking is also 26 mm. in diameter and thin-lined. Here the *IX* measures 15–16 mm. wide.

Type d is 26 mm. in diameter, but the circle is thick-lined (about 1 mm.). This type is easily distinguishable because the *5* is directly beneath the *I*, rather than being to the right of it ... as is the case with the other four types.

Type e is also 26 mm. in diameter and thick-lined. The *5* is to the right of the main stem of the *I*.

I'll restate the rules for identifying the types of the circular postoffice designator another way. If the circle is small, it's type *a*. If the *IX* is 13–14 mm. wide, it's type *b*. If the *5* is directly beneath the *I*, it's type *d*. Type *c* and *e* are only distinguishable by the thickness of the circle, type *c* is thin-lined, type *e* is thick-lined. This is likely to cause an identification problem only for a very weak strike.

The *Fliegerpost Przemysl* handstamp was applied in black, blue, or lilac ink, usually in the upper center of the postcard. The first and most common variety reads *Fliegerpost Przemysl / Jänner 1915* which translates "Aviator Mail Przemysl / January 1915." This was shown in Figure 3–12. Figure 3–13 shows the other five types without reduction.

All six are unframed two-line markings. The first line always reads *Fliegerpost Przemysl.* The date in the second line is always *1915.* Tranmer notes obvious forgeries that used the date *1914.* The identifying characteristics of each type are provided below.

Type 1 -- as previously noted -- is 67 mm long and the second line reads *Jänner 1915.* [Note, the second line includes the period.] It is the most common of these markings. Its identification is immediately obvious. It is the only type to include the word *Jänner* that, by the way, means "January" in the Austrian dialect of German. It appears that this marking was used only on the January 18 flights, and that none of the other five types were used on that date.

Type 2 is the same handstamp as Type 1, but was applied after cutting out the word *Jänner*. Therefore, the second line contains only *1915.* and it is not centered. Rather, it is oriented toward the right under the letters "*rzem*" of *Przemysl*.

Type 3 is also 67 mm. long. The font is the same as types 1 and 2, but the *1915.* is centered in the second line. There is a period at the end of the second line, but it often is not readily apparent.

Type 4 is the same as type 3 in every way except that the second line (*1915.*) slopes downward from the orientation of the first line. The period in the second line is again not always visible. Distinguishing between types 3 and 4 presents the greatest difficulty among the *Fliegerpost Przemysl* markings, but with a clear strike is usually not a problem.

Figure 3–13 Types of *Fliegerpost Przemysl* Marking -- Types 2, 3, 4, 5, 6 (top to bottom) -- All with *IX/54* Circular Marking Type "d"

Type 5 uses a different font than the previous types. The first line is 59 mm. long. The *1915* in the second line is centered and there is no period. The letters in the word *Fliegerpost* are set crookedly, but this is of no consequence to identification. The length of the first line says it all.

Type 6 is virtually identical to Type 5 except its length is 57 mm. Therefore, if the strike is complete you should have no difficulty distinguishing between these two types.

Figure 3–13 shows types 2 through 6 without reduction to assist you with identifications.

Most collectors of Przemysl airmail try to collect combinations of these two basic markings. To simplify this approach, I use the combination of a number (1–6) and a lower-case letter (a–e) to identify every postcard. So, for instance, I refer to the most common combination seen at the top of Figure 3–12 as *Type 1a*. Remember, all the combinations do *not* exist.

Many authors have tried to estimate the scarcity of the various combinations, and they are generally in agreement. But, in their writing they have used point systems or various currencies to assign the values. I have taken their numbers and normalized them. In other words, I have arbitrarily assigned the most common varieties, types 1a and 1d, the value of *1.0*. All the other types that exist are assigned values which are multiples of this base. So the values approximately indicate their relative scarcity.

I must note that the scarcity of the combined marking types is only one of the many factors that establish the value of a Przemysl airmail card. Therefore, the multiples shown in the table are *not* an indication that a card should sell for that multiple of the going-rate for the commonest card. Combinations that do not exist are indicated by two dashes.

Scarcity of Przemysl Airmail Marking Combinations

	a	b	c	d	e
1	1.0	1.1	2.0	1.0	1.2
2	10.0	1.2	2.0	1.2	1.7
3	--	2.5	--	1.7	--
4	--	2.3	--	1.4	6.6
5	6.6	3.3	10.0	1.7	3.3
6	1.5	1.3	2.7	1.1	2.7

My final point concerns suggestions on how to collect the types of the regular Przemysl airmails. An obvious approach is to go after an example of every existing combination. While a worthy goal, it represents a difficult and very expensive long–term task. A less arduous approach (and one that puts you on a path to the above goal) is to collect a group of cards that show an example of every marking type. Many years ago I decided to follow the latter path. In deciding which cards to pursue, the thought immediately came to mind to minimize both cost and the number of pieces involved. A careful analysis shows that collecting only six cards -- 1e, 2c, 3b, 4d, 5d, and 6a -- satisfies this goal. The sum of their scarcity ratios from the table above is 10.3.

When I saw many cards offered with the Type 1 *Fliegerpost Przemysl* marking and the Type d *IX/54* circular marking, I realized there might be a better way. So, instead of going after the minimum six cards, I pursued all the combinations in the first row and fourth column of the table. Though there were ten cards involved now, the sum of their scarcities was still only 13.4. With years of effort, I succeeded with this approach as you see from Figures 3–12 and 3–13. I also found it to be a clear and concise way to exhibit the marking types. And, to this day I still haven't seen an example of type 3b addressed to Hungary!

Other Markings and Varieties. As mentioned above, two other markings are found on *every* regular Przemysl airmail card. These are numbering and censorship markings.

Every card was numbered. This was done either manually, or with one of two *numerators* (machines that automatically increment the number printed each time the device is used). The numbering was done in the upper left hand corner of the cards.

The manuscript numbers appear to go up to about 30,000. Most are in pencil; some are in ink. Most cards with manuscript numbering were flown on January 18 but, a few can be found on flights as late as February 12.

The machine numbering began on January 11. Beginning with the February 18 flight, only cards numbered by machine were flown. The stamped numbers are in black ink. Two numerators were used. The first printed numbers that are about 6 mm. in height. Sobetzky is more precise, and measures these numbers as 5¾ mm. high. Cards are found with these numbers up to about 35,000. The second numerator printed numbers about 6½ mm. in height. These are found with numbers up to about 18,000.

Censorship was a two–step process. An officer in each unit read and signed each card. Typically this was done with the word *Uberprüft*, *Zensuriert*, or something similar followed by the officer's name and rank. Some wrote only the letter "Z" and their signature. Other officers had handstamps with the word "Zensuriert" that they applied with their signature. (Don't confuse these private censorship handstamps with the official ones discussed later.)

Figure 3–14 Przemysl Airmail with Six Markings

The manuscript censorship indications were usually in pencil, sometimes in ink. More often than not, these look rather like a scribble near the lower left corner of the card. Don't be too surprised if this marking is missing. Some collectors with more good intentions than knowledge have erased these "scribbles" to clean–up the card.

The second step of censorship was the check at headquarters that an officer had signed each card, and the application of a *ZENSURIERT.* handstamp. These handstamps were 48–52 mm. long with ink in red shades ranging from pink to deep lilac. Usually this mark was applied in the middle of the card, often on top of the address. It can be seen on several of the cards in Figures 3–12 and 3–13.

In some cases, rather than the *ZENSURIERT* handstamp, a handstamp reading *Zur Beförderung geeignet* was used. This means "suitable for forwarding." These exist in several types. They are much scarcer than the normal *Zensuriert* marking. According to Sobetzky, they are found on less than 5% of the Przemysl airmails. Figure 3–12 (types 1b and 1d) show examples of these markings.

Occasionally, two other markings are found on regular Przemysl airmail cards -- though technically they shouldn't be there. These are the unit marking of the sender's unit and the standard double–ring circular date stamp of the Przemysl postoffice. Therefore, the maximum number of postal markings on any Przemysl airmail card is six. Figure 3–14 shows an example with all these markings.

Another occasional oddity on Przemysl airmail cards is stamps! Obviously they served no valid postal need in the fieldpost system. As surrender approached, some soldiers believed that stamps might help get the mail delivered should the cards be captured by the Russians. So, they gave it a try. Both Austrian and Hungarian stamps were used, sometimes in combination. When stamps are present, they are usually canceled and tied by the *Zensuriert* handstamp. Stamped Przemysl mail is quite scarce.

Several authorities have mentioned the use, on the last flight, of "strips" rather than cards. Supposedly, cards were cut into five parts to increase the number of messages per unit of weight. The only example I have seen could easily have been "created" from an unused card since it has a horizontal orientation rather than the more logical vertical orientation. I certainly can't say that this variety of Przemysl airmail doesn't exist, but I can suggest the old Latin maxim *caveat emptor* -- "buyer beware."

If you see many Przemysl cards addressed to Hungary, you may notice that two areas predominate as destinations. These are the heavily populated Budapest area, and areas of northern Hungary geographically adjacent to Galicia. Neither of these should come as a surprise. It does mean, though, that Przemysl mail addressed to remote areas of southern and eastern Hungary is very scarce. This fact, while of little concern to Hungarian airmail collectors, is of considerable importance to Yugoslav and Romanian airmail collectors. They often view Przemysl cards to those areas as "forerunners" in their specialties.

Personal Mail of Gen. Kuzmanek. Of interest to many collectors ... though few will ever own an example ... is the personal mail of General Kuzmanek. Clearly such mail is very rare. According to Tranmer only three pieces exist. If that is the case, they are the cards to Austria illustrated by Clement and Tranmer, and the one shown in Figure 3–15 that is addressed to Hungary. Some of the facts associated with this mail are indisputable. On one point, I strongly disagree with the "authoritics."

The Kuzmanek postcards are not flimsies. They are standard-weight fieldpost cards. They bear the very rare combination marking type 4e, as well as the circular unit mark of *K.u.k. Festungskommando in Przemysl*. They were neither numbered nor censored, though they show the red *Zensuriert* handstamp. On their reverse they have a rectangular purple handstamp found nowhere else. Translated it reads: "Delayed by the second encirclement of Przemysl and the resultant interruption of postal service." Figure 3–15 shows both the front and reverse of a Kuzmanek card to Hungary.

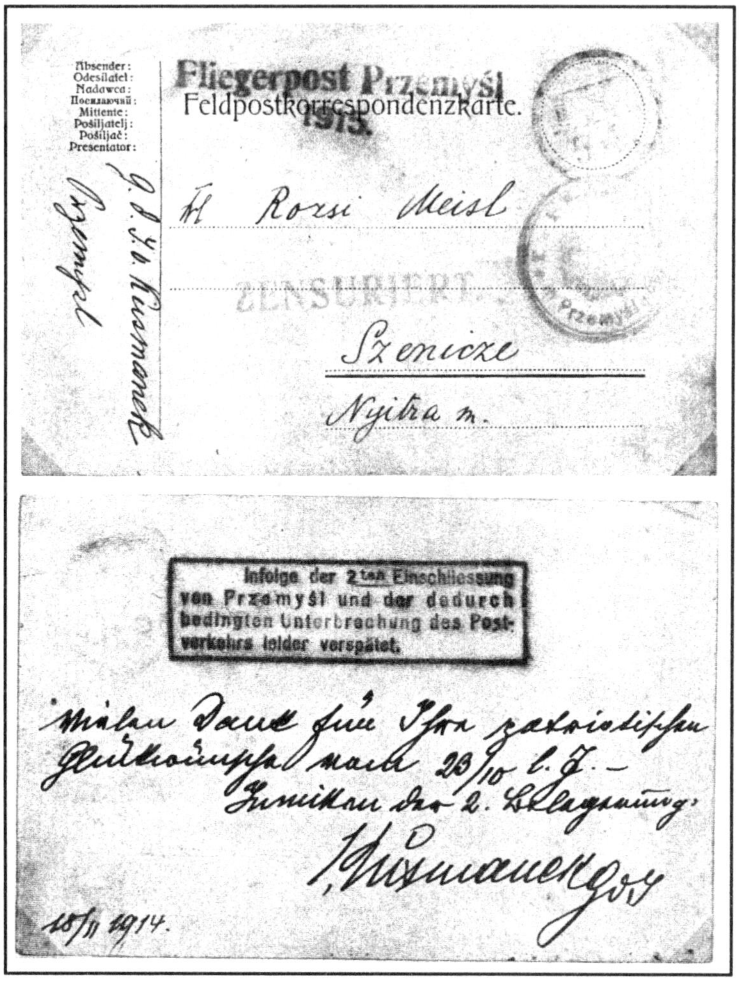

Figure 3–15 Personal Airmail of Gen. Kuzmanek

The cards have a manuscript message (in German) and the bold signature of "Kuzmanek GoI." At the lower left of all three cards is the date "18/11 1914." Tranmer interprets the date -- November 18 -- as proof that the cards were flown on the forerunner flight of November 23. I believe this is a bad assumption! It requires that not only the regular airmail service was anticipated, but that the markings ... even the type 4 marking not yet found on January mail ... were prepared months in advance.

I believe there is a scenario that makes a lot more sense. Kuzmanek wrote the cards in November, put them aside, and forgot about them. When he or an aide cleaned out his office prior to the surrender, the cards were found. The cards were hastily handstamped with the special "delayed" marking (as befits the commanding general) and included with the last airmail out.

Balloonpost of Przemysl. Based at Przemysl were the three balloon detachments of the 3rd Fortress Artillery Regiment of Prince Kinsky. Their primary duties were artillery spotting using tethered kite balloons. The 1st Balloon Detachment, under the command of Lt. Col. Julius Nimmerrichter, also employed five free balloons in reconnaissance missions.

On January 12, 1915 the announcement of the organization of a balloonpost, to assist the troops in corresponding with their relatives, was made in **Tábori Újság**. Nimmerrichter was organizer and head of the balloonpost.

10,000 lightweight cards were prepared. They were similar to those used in the regular airmail, except that their color was blue or blue–grey. Likewise, they were sold for 50 fillér with proceeds going to the widows and orphans trust fund. Unlike the regular airmail, only the number of cards was sold that could be carried by each balloon under construction.

The plan was to use unmanned balloons at times of favorable wind and weather. The balloon apparatus stood about two meters high. They were constructed of painted wrapping paper and had a gas capacity of eight cubic meters. Suspended below the gas bag were the mail packet and a cone for sand ballast. Figure 3–16 shows this type of balloon ready for release.

The balloons themselves weighed about two kilograms, and they carried about three kilograms of sand ballast. The sand would be gradually released in flight through an opening at the point of the cone. The balloons could accommodate about two kilograms of mail ... 500–600 cards ... including the outer wrappers.

A flying time of 5–6 hours was planned. This would take the balloons more than 120 kilometers ... further if a north wind carried them over the Carpathians.

The mail was wrapped and double–gummed to create a waterproof packet. A printed label was attached to the packet. In all the languages of the Empire, it asked that the finder turn over the packet to any government office for a 10 Korona reward.

The rules and postal markings were similar to those for the regular airmail. The cards were numbered with a 6 mm. numerator. Numbers up to about 6,000 exist. A 55 mm. long two–line *Ballonpost Przemysl / 1915* marking was applied. The cards in Figures 3–17 and 3–18 show this marking. Also, an *IX/54* circular postoffice designator was applied. This was *Type a* in most cases, but types *c* and *e* are found on rare occasions (see Figure 3–17). The three–line unit marking of one of the balloon detachments is often seen at the left side of the cards. The sale of cards was limited to one per officer, and one per fifteen enlisted men.

Figure 3–16 Unmanned Balloon at Przemysl

In all, ten balloons were prepared. Five of these were released on January 20, three on March 7, and two, which are known to have been captured by the Russians, were released later. Of the eight balloons that may have survived, the fate of only three is certain. All three of those were found in Hungary.

The first balloon found was from the group released in March. It was found in Turja-Remete (now Turji Remeti, Ukraine) on March 19. The mail packet was passed to FPO 47 which postmarked each card that day. Figure 3–17 shows a card from that balloon that was delivered to Ecser on the 26th.

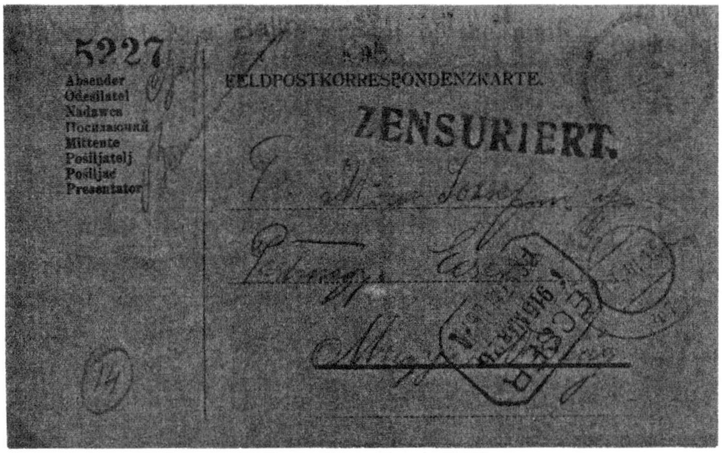

Figure 3–17 Balloonpost to Turja-Remete

The second and third balloons found are believed to be from the group released in January. The second was found near Lemes (now Lemesany, Slovakia) on May 18. This mail was postmarked there. The third was found at Margitfalva (now Margecany, Slovakia) on June 20 and the mail was postmarked there.

The fate of the other five balloons is unknown. Probably much of their mail is lost; but, some mail from those balloons is known. Certainly, this mail may have fallen into hands that moved it into the philatelic market after the war. That is possibly the case for the two balloons captured by the Russians as well. A packet may have even been found years later. If this mail is actually from the lost balloons, then it was flown and is of definite interest to aerophilatelists. On the other hand, like the regular airmail cards, a few unused balloonpost cards are known and could be used to create rarities. Again, buyers beware!

Finally, let me tell you why I am thoroughly convinced that at least one of the "lost" balloons was found in 1915, and its mail was forwarded to addressees through the postal system. The card shown in Figure 3–18 is dated January 13, and so is certainly from one of the January balloons. It does not bear a postmark of northern Hungary, where it must have been found. But, it has a June 16 receiving mark of Szolnok ... the city to which it is addressed. It was in the Hungarian postal system in 1915.

The final chapter regarding Przemysl balloonpost was written in the four days before surrender. The plan was that the five reconnaissance balloons escape, with two or three officers each, official documents and some mail. The mail was a mixture of *Fliegerpost* and *Ballonpost* cards.

Figure 3–18 Card from a "Lost" Balloon

The orders were, if captured, to destroy both the documents and the mail. The first balloon to leave, on March 19, was the *Przemysl*, followed by the *Austria*, the *Erzh. Josef Ferdinand*, and the *Schicht*. The *Stiermark* broke loose of its moorings and was carried off unmanned by the wind. Russian artillery brought it down.

The four manned balloons that ascended did no better. All landed in Russian controlled territory. Three of the four balloon pilots followed orders and destroyed everything. The other, the *Austria* with Lt. Col. Nimmerrichter aboard, destroyed the documents but not the mail. A few pieces of this mail survive, with markings of Russian military censors and the Red Cross. They were apparently forwarded through the International Red Cross office in Copenhagen.

There is a problem with the identification of this airmail. Generally, it is not distinguishable from the mail recovered from the March 22 flight of Sgt. Meltsch. Only if the mail is dated after the 19th, can it firmly be attributed to the Meltsch flight.

Figure 3–19 Manned Balloon at Przemysl

Not many of us will have that problem though. Very few pieces of this mail have been found. They are very rare. I have never seen a piece addressed to Hungary, though that certainly doesn't mean that none exists. Figure 3–19 shows an aeronaut in the basket of one of the Przemysl reconnaissance balloons before ascent.

In–Bound Przemysl Mail. The flight into Przemysl during the first siege did not carry mail. During the second siege, six planes are known to have carried mail into Przemysl and seven others may have. That would seem to indicate such mail would be as common as outbound mail. That certainly is not the case!

Most items claimed to be "Inbound Przemysl Airmail" are outright fakes. Any guidelines for identifying genuine examples of such mail are necessarily very general. For example, the mail must be postmarked during the siege and addressed to an FPO or unit known to be in Przemysl. Beyond that, there's not much to go on.

There were, though, many problems related to the survival and identification of inbound mail. Unlike the outbound mail, most of it had no clear postal markings that associated it with the siege. Second, prior to surrender the entire garrison was ordered to destroy all their personal papers before turning themselves over to the Russians. Probably a few did not comply with this order. Even then, the fragile pieces of mail had to survive the rigors of years in a POW camp. Considering all that, it is not surprising that little identifiable inbound Przemysl airmail survives today.

The commonest material claimed to be inbound Przemysl airmail involve Fieldpost cards imprinted with the 45x27 mm. rectangular handstamp shown in Figure 3–20. This marking was created for inbound mail by one of the fieldpost offices (56 or 101) attached to 4th Army headquarters at Brzesko. It was sold to benefit charity in Vienna and elsewhere.

Unused examples, such as the one shown, are common. How hard would it be to add the

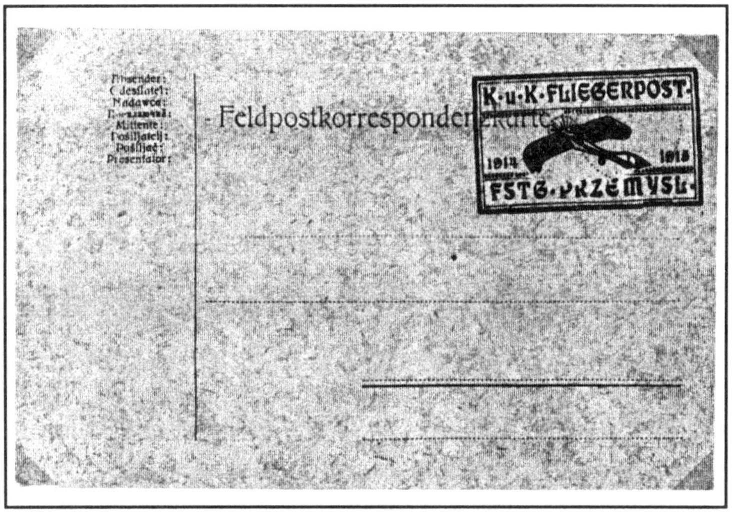

Figure 3–20 Brzesko Inbound Airmail Handstamp

address of a fieldpost office known to be in Przemysl, and to add an innocuous message on the reverse? I feel this is how many of the pieces of "inbound airmail" originated. It is an interesting aside that this handstamp was preserved and, after the war, was used to "cancel" cheap stamps on a "piece." These flooded the 1920's philatelic market and are still quite common.

Figure 3–21 Post–War Philatelic Creation

An example is shown in Figure 3–21.

You've now been told the story of how a fair amount of airmail was flown into the besieged fortress, and why so very little has survived. Obviously, more mail was sent to Przemysl by loved ones than could possibly be flown there.

This backlog of mail accumulated until the surrender of Przemysl. Only then, each piece was returned to the sender as undeliverable. Such mail, since it was not flown, is of limited interest to many aerophilatelists. I have included an example in Figure 3–22 because of the interesting red bilingual handstamp it bears. That handstamp reads, "*Returned to Sender / Undeliverable because of the War Situation*" in both Hungarian and German. The cover was folded to show the Hungarian stamp and postmark on the reverse.

Thus ends the story of what many consider to be the greatest airmail exploit of World War I.

Seaplane Bases on the Adriatic

Lining the eastern shore of the Adriatic are many towns that to this day are rather isolated. In the World War I era, roads through the rugged coastal mountains were not much more than "goat trails." There were few places that railroads penetrated to the coast. As a result, most communications and transport were limited to ships, and they are very slow.

The Austro–Hungarian navy had bases up and down the Adriatic coast. Fast communication between them took on utmost importance in

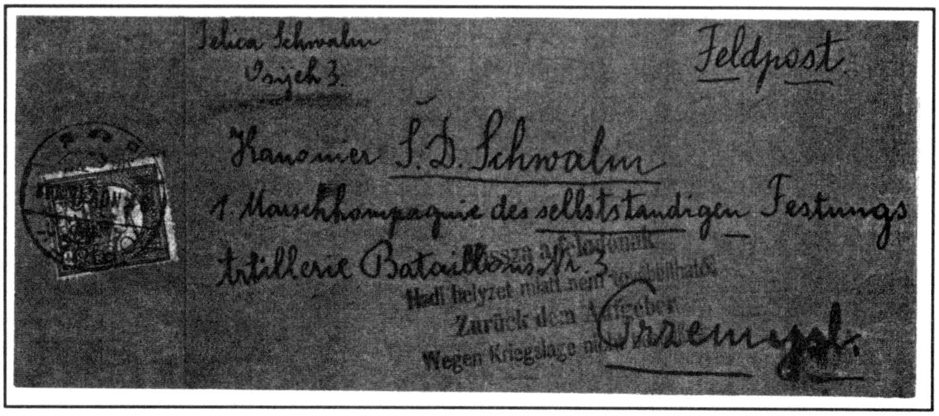

Figure 3–22 Inbound Przemysl Mail –– Returned to Sender

1915 when Italy came into the war on the Allied side. To satisfy this, and the military objectives, the Navy established a string of seaplane stations from the Italian border to Albania. The Seaplane Corps performed and supported both offensive and defensive military missions. Of particular concern was defense of the large naval base on the Bay of Cattaro (also a major German submarine base), and of ship convoys in the Adriatic sea lanes.

Figure 3–23 Navy Flying Boat in Pola Harbor

Background. The navy first acquired a landplane in 1910. It was based at Pola. The Seaplane Corps began when the navy's first flying boat was tested at St. Catarina, within the massive harbor of Pola that served as naval headquarters. That was late in 1911. At the end of 1913, just months before the war began, the Seaplane Corps numbered four aircraft and about 150 officers and men. It was still confined to St. Catarina. By then the navy had decided on Lohner flying boats for the seaplane corps, and these were the mainstay of seaplane operations until the war's end. Figure 3–23 shows a Navy Lohner flying boat in Pola harbor.

Toward the end of the war, the Seaplane Corps was operating 266 flying boats, with a staff of several thousand officers and men, at 15 distinct locations. They provided air superiority over the Adriatic throughout the war. This was the case despite operating against both the Italian air corps and many units of the world–class French air force. Their solid performance was largely the result of the quality of training received at Cosada, the naval flying school on an island outside Pola harbor.

Courier Mail. In the early years of the war, before most of the seaplane bases were operational, the Navy operated small land planes to provide courier services between the various naval bases on the Adriatic. Later, the Seaplane Corps assumed much of this task. Some Navy land–based planes continued to fly.

The literature mentions three courier mail handstamps that were applied to mail carried on these planes. They read: *Kurierdienst / Eindecker Pola*, *Kurierdienst Ikarus*, and *Mit Küstenflugzeug / Schwalbe / Befördert*. I have found a fourth that reads: *Mit Küstenflugzeug M4 / Befördert*. The M4 was a land plane operated by the Seaplane Corps'

Figure 3–24 Kumbor–Fiume Courier Mail

Figure 3–25 Map of Seaplane Bases

administrative unit (*Seeflugleiturng*) until at least October of 1916. Figure 3–24 shows a card with this "M4" marking, though it is in extremely faint red ink. The card is from the collection of Sobetzky, who identified it as having been flown on a February 17, 1916 courier flight between Kumbor and Fiume. Material with any of these courier handstamps is exceedingly rare, and those with a Hungarian connection even rarer.

Two other major courier services involving Navy facilities were of importance later in the war. The first of these was an air courier service operated by the Navy between Pola and Udine, the Army's headquarters on the Southern Front. This service connected with the Army's courier service between Udine and Vienna. It provided the fastest means for sending data between Pola and Vienna. Mail carried on this route was overwhelmingly military dispatches, and little if anything identifiable is in private hands.

The German Navy operated the second of the 1917–18 courier services between Pola and the naval base at Cattaro. They carried mail to and from German submariners in port at Cattaro. This flown mail is very rare; but in any case, the likelihood of any of it being of interest to Hungarian aerophilatelists is low.

Seaplane Stations. The naval facilities that housed seaplane stations changed from time–to–time during the war. The map shown in Figure 3–25 identifies the general location of all the seaplane bases, whether or not postal markings of the Seaplane Corps exist. Below, I discuss each base and its markings. Arbitrarily, I have decided to deal with the seaplane stations from north to south.

It is the postal markings of these seaplane bases that we are primarily concerned with here. I cannot illustrate every one. The examples in Figure 3–26 show the five general forms of the markings. In the lists of known markings for each base, I identify the general form of each. Then I provide any particulars described by my references. The five most common forms are:

1) Double–circle: Durz
2) Single–circle: Pola
3) Boxed: Fiume (FPO 406)
4) One–Line: Curzola
5) Two–Line: Sebenico.

Note that the Durz marking includes the *dual arms* of the Empire, and that the Pola marking shows the Habsburg double–headed *eagle*. I will use those words in my descriptions.

Figure 3–26 Forms of Postal Markings

The lists of known seaplane postal markings are included with my discussion of each seaplane base at the end of this chapter. The markings listed represent a composite of those recorded by Clement, Sobetzky, and my own research. An asterisk (*) marks my new findings.

Was it flown? It is important to remember that all the markings we are discussing here are "unit markings" –– or in some cases combined unit and censorship markings. As with the Army aviation unit markings, they do *not* indicate that the mail was flown. Unit markings were required on personal mail to get the "free franking" privilege of active–duty military personnel. It is only the very rare courier markings mentioned above that explicitly establish that the mail was flown.

But, the Navy's seaplane bases were different than the Army units that all had fieldpost offices trailing not far behind and railroads close by. Mail service in the Adriatic was extremely slow. So, it is natural that pilots expedited mail by flying it to a base with good railroad connections. In the north, railroads reached most of the mainland naval bases. Trieste and Fiume were major commercial ports and rail centers. Pola's military importance supported a connector line to Trieste. In the south, railroads were practically non–existent. Gravosa was on a small rail line that terminated at Zelenika, on the north shore of the Bay of Cattaro. That small railroad wound through the mountains to Sarajevo. That was it. The remainder of the mainland naval bases, and all the island bases were very isolated.

So, we can expect that a small but significant percentage of the mail of Seaplane Corps personnel was flown somewhere enroute. It's there, but it's difficult if not impossible to identify. Here are some clues.

(1) Does it have postal markings of two Adriatic cities on the same day? Probably flown. (2) Does it have a unit marking of one seaplane station and a postal cancellation from another base (most likely the *Marinepostamt* in Pola, or the civilian postoffices at *Teodo* or *Zelinika* near Kumbor). Probably flown. (3) Is it a picture postcard of one city with a seaplane station, and the postal markings of another? Possibly flown.

The important thing in making these decisions about how to represent a piece of mail (or whether to buy it) is to use common sense, and don't kid yourself. If you have real doubts in your mind about a piece being flown, it probably wasn't! Yet, it still may be very collectable.

That brings us to Fregattenleutnant Oskar Buchberger of Torpedo Boat 85F. He must have been a philatelist because postcards addressed to him with the unit markings of every seaplane base in operation exist ... plus a lot of other naval material as well. In some cases, apparently, only Buchberger material exists. It's clean, almost pristine. Too clean; the markings look to be all canceled to order. They bring a good price, but not what real mail with these markings brings. Probably most of my readers can be unconcerned ... I haven't seen a Buchberger piece yet with a Hungarian connection.

Figure 3–27 shows how the Navy extended the range of the flying boats for bombing missions against Italy. In the WWI version of an aircraft carrier, flying boats were winched between the sea and the decks of naval ships.

Figure 3–27 World War I Aircraft Carrier

List of Seaplane Bases and their Markings.

Please remember, the markings listed here reflect material reported by Sobetzky in the 1930's, Clement in the 1950's, and my research over the last quarter of a century. I have found several markings, not reported in the earlier work. These are marked by an asterisk (*). There are still probably others out there. Just because it's not listed, don't assume that it doesn't exist. We all still have a lot to learn in this area.

Also, sometimes it's not only the mail of the seaplane units that is of interest to Hungarian aerophilatelists. The mail of other naval personnel, at bases with seaplane facilities, may also have been flown. For example, Figure 3–28 shows a cover with the unit marking of the headquarters staff at the Fiume naval base. It also shows postmarks of Fiume and Pola on the same day. Quite possibly, it was flown.

Figure 3–28 Cover Probably Flown Fiume–Pola

Grado (now Grado, Italy). The seaplane station at Grado was established in November 1917. It functioned under the direction of the base at Trieste, and was intended primarily for defensive operations. No postal markings of the Seaplane Station at Grado are known.

But, that may not be the end of the story. I have seen two postcards with an unusual marking. They are from the same correspondent and addressed to Prague, so there is no Hungarian connection. That's why they aren't illustrated. Pola seaplane station unit markings and naval postoffice postmarks are on both. Both also have a previously unreported 39 mm. violet straight–line handstamp that reads *Fliegerstation No 6*. Neither reading, nor the experts I've contacted, have provided any clue to the source of that mark. The Army is not known to have used the designation *Fliegerstation*, and note that it is *not* preceded by the ubiquitous *Kuk*. One of the cards pictures a beach scene at Grado.

How often have you sent home a picture postcard from a place you weren't visiting? Was the seaplane base at Grado called *Aviator Station 6* for some reason? Possibly both cards were flown. More likely, that station is elsewhere. Still, one card may have been flown from Grado to Pola. Was *Aviator Station 6* one of the Navy's land–based airfields in the Pola area? Valbandon, perhaps? Then it probably wasn't flown. Or was it an Army airfield elsewhere, perhaps even in Bohemia? Was *Aviator Station 6* an Austro–Hungarian facility at all? I don't have answers yet. These are the kind of questions you get into. As I already said, there's a lot still to be learned and new postal markings to be found and studied.

Triest (now Trieste, Italy). Trieste was the major commercial seaport of Austria. It also was home to extensive naval facilities. The seaplane station was established in June 1915 when Italy joined the war on the Allied side. It was the principal station in the north and responsible for operations at Grado and Parenzo. Only one Seaplane Corps postal marking is known.

Trieste Seaplane Station #1 -- One–line; *Kuk Seeflugstation Triest*; 65x4 mm; red.

Parenzo (now Porec, Croatia). The seaplane station at Parenzo was established in September 1916 to protect the vulnerable west coast of Istria from Italian attack. It operated initially under the direction of the base at Trieste. Only one Seaplane Corps postal marking is known.

Parenzo Seaplane Station #1 -- Single-circle with dual arms; *Kuk Seeflugstation Parenzo*; 29 mm diameter; red.

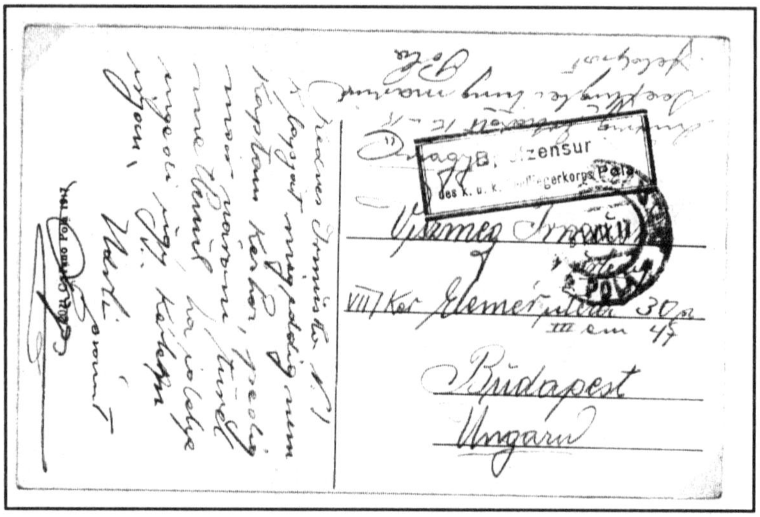

Figure 3-29 Pola Seaplane Corps Headquarters Marking

Pola (now Pula, Croatia). The origins of the seaplane station at Pola have already been given. In this section, I deal with five seaplane units in the Pola area. They are: (1) *Kuk Kommando des Seeflieger Corps* (headquarters staff for the entire Seaplane Corps); (2) *Kuk Seeflugleitung* (administrative staff for the entire Seaplane Corps); (3) *Seeflugstation Pola* (Pola's seaplane base, initially at St. Catarina within the harbor; later at Puntisella, just north of the entrance to Pola harbor); (4) *Schulflugstation Cosada* (the flying school on an island outside Pola harbor); and (5) *Seeflugstation Puntisella* (just north of the harbor entrance; in 1917-8, Pola's only seaplane base).

In addition to normal unit markings, at Pola, Cosada and a few other bases, combined unit and censor marks were used. These are listed following the circular and straight-line unit markings.

Pola Headquarters #1 -- Single-circle with dual arms; *Kuk Kommando des Seefliegerkorps;* 25 mm diameter; red or black.

Pola Headquarters #2 -- One-line; *Kuk Kommando des Seefliegerkorps*; 86x3 mm; red.

Pola Headquarters #3 -- Boxed; *Briefzensur / des Kuk Seefliegerkorps Pola*; 46x15 mm; red or black*. (see Fig. 3-29)

Note: The illustrations that follow were selected to show the variety of the material available and a couple of unusual pieces.

Pola Administration #1 -- Single-circle with dual arms; 24 mm diameter; *Kuk Seeflugleitung*; red.

Pola Administration #2 -- Single-circle with dual arms; 29 mm diameter; *Kuk Seeflugleitung*; red*. (See note a)

Pola Administration #3 -- Single-circle with dual arms; 39 mm diameter; *Kuk Seeflugleitung*; red. (See note a)

Pola Administration #4 -- One-line; *Kuk Seeflugleitung*; 44x4 mm; red or violet.

Pola Administration #5 -- One-line; *Kuk Seeflugleitung*; 51x4 mm; red.

Pola Administration #6 -- Boxed; *Briefzensur / der Seeflugleitung Pola*; 50x14 mm; red. (See note b)

Notes: (a) I included #3 for completeness; but, I suspect a typographical error in Clement and that it is the same marking as #2.

(b) #6 above is reported by Sobetzky with the letters "*leitung Pola*" missing; probably just a light strike.

(c) The *Seeflugleitung* were the folks who held the money. Therefore, it is much more likely that you will find their handstamps on financial documents, such as receipts and expense reports, than on mail.

Pola Seaplane Station #1 -- Single-circle with eagle; *Kuk Seeflugstation / Pola*; 20 mm diameter; red. (see Fig. 3-30)

Figure 3-30 Pola Seaplane Station Marking

Pola Seaplane Station #2 -- Single-circle with eagle; *Kuk Seeflugstation Pola*; 25 mm diameter; red. (see Fig. 3-26)

Pola Seaplane Station #3 -- Single-circle with eagle; *Kuk Seeflugstation Pola*; 37 mm diameter; red.

Pola Seaplane Station #4 -- One-line; *Kuk Seeflugstation Pola*; 63x3 mm; red.

Pola Seaplane Station #5 -- One-line; *Kuk Seeflugstation Pola*; 68x4 mm; violet.

Pola Seaplane Station #6 -- Boxed; *Briefzensur / der Kuk Seeflugstation Pola*; 50x14 mm; red.

Pola Seaplane Station #7 -- Boxed; *Briefzensur / der Kuk Seeflugstation Pola*; 53x15 mm; red or violet.

Note: #1 and #7 above are the commonest of all the seaplane base postal markings. Yet, they are not *all* that common ... especially on mail to Hungary.

Cosada Flying School #1 -- Single circle with eagle; *Kuk Schulflug-station / Cosada*; 30 mm diameter; red.

Cosada Flying School #2 -- One-line; *Kuk Schulflugstation Cosada*; 63x4 mm; violet*. (see Fig. 3-31)

Cosada Flying School #3 -- Boxed; *Zensuriert / Kuk Schul-flugstation*; 44x14 mm; violet.

Cosada Flying School #4 -- Boxed; *Briefzensur / der Kuk Schulflugstation*; 57x19 mm; violet. (See note)

Cosada Flying School #5 -- Boxed; *Briefzensur / der Kuk Schulflugstation Cosada*; 57x19 mm; violet*.

Note: I have my doubts that #4 exists, since the size is identical to #5. But, both Clement and Sobetzky list the marking as "without the town name." So, I included it for completeness.

Puntisella Seaplane Station #1 -- Single-circle with eagle; *Kuk Seeflugstation Puntisella*; 27 mm diameter; red.

Note: The seaplane base at Puntisella operated under its own name in 1916 while St. Catarina was still in operation. Later, the unit at Puntisella was simply known as *Seeflugstation Pola*. This leads to confusion since, for some time, the two designations were used interchangeably.

Fiume (now Rijeka, Croatia). Fiume was the major commercial seaport of Hungary, and home of important naval facilities. The seaplane station was established in August 1916, and its mission was reconnaissance in the

Figure 3-31 Previously Unrecorded Cosada Marking

Gulf of Fiume. It operated under the direction of the station at Pola. Only two flying boats were stationed here. As the only seaplane station in Hungary, it is very important to Hungarian aerophilatelists. Unfortunately, there is only one postal marking of the seaplane base at Fiume; and, it is generally described as the rarest of all the seaplane postal markings.

Figure 3–32 Fiume Seaplane Base Marking on 1917 Cover

Fiume Seaplane Station #1 -- Boxed; *Briefzensur des / Kuk Seeflugstützpunktes / Feldpost 406*; 66x27 mm; red. (see Figs. 3–26 and 3–32)

Notes: (a) Sobetzky mentions that this marking was used on courier mail between Fiume and Abbazia (a naval facility on the Austrian side of the Gulf of Fiume; now Opatija, Croatia) in August and September 1917. He notes that only six examples exist.

(b) FPO 406 is an army fieldpost unit operating in Fiume. This is one of only two direct postal tie–ins between the Army and the Seaplane Corps postal systems.

(c) When the Sobetzky collection was sold, it contained a cover from Fiume with the same wording as #1 above; but, it appeared to be typewritten in red. I have chosen not to include it in the listing since it is probably a unique item, and not a normal postal marking.

Lussin (an island; now Losinj, Croatia). The seaplane base on Lussin was established in June 1916. It was to provide support in both offensive and defensive missions in the sea lanes adjacent to the north Dalmatian islands. It was under the direction of the station at Pola. Two postal markings are recorded for the seaplane base on Lussin.

Lussin Seaplane Station #1 -- One–line; *Kuk Seeflugstützpunkt Lussin*; 78x4½ mm; violet (?).

Lussin Seaplane Station #2 -- Three–line; *Briefzensur des / Kuk Seeflugstützpunktes / Lussin*; 45x13½ mm; violet.

Zara (now Zadar, Croatia). The seaplane base at Zara opened in November 1917. It was a minor support facility. No postal markings of this base are known.

Sebenico (now Sibenik, Croatia). Construction of the seaplane facilities at Sebenico began in June 1914. It was officially designated a *Küstenflugstation* (Coastal Flying Station) in November of that year, and was designated a seaplane station in August 1916. Its missions were mine and submarine defense in the bay,

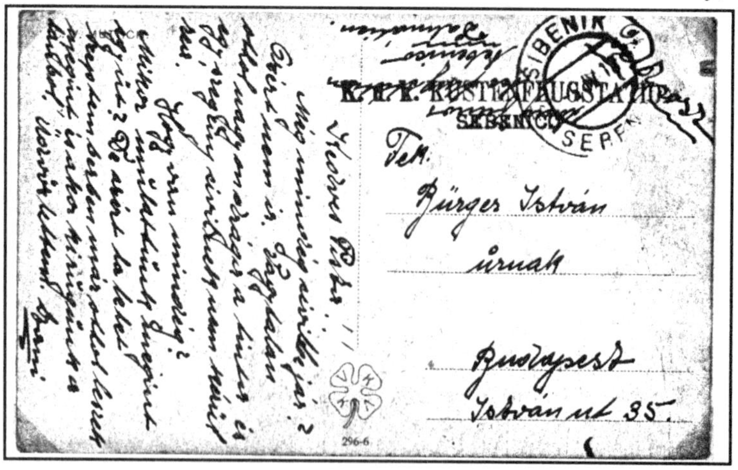

Figure 3–33 Seaplane Station at Sebenico Mail

and long–range reconnaissance of a portion of the Italian coast. Three Sebenico postal markings are known.

Sebenico Seaplane Station #1 -- Single-circle with eagle; *Kuk Küstenflugstation Sebenico*; 24 mm diameter; red or violet.

Sebenico Seaplane Station #2 -- Two-line; *Kuk Küstenflugstation / Sebenico*; red or violet. (See Fig 3-26 and 3-33)

Sebenico Seaplane Station #3 -- One-line; *Kuk Seeflugstation Sebenico*; 72½x4 mm; red.

Rogoznica (now Rogosnica, Croatia). The seaplane base at Rogoznica was established in June 1916 as defensive facility under the direction of the station at Sebenico. It had responsibility for the protection of sea lanes on a portion of the Adriatic coast. One postal marking is recorded.

Rogoznica Seaplane Station #1 -- One-line; *Kuk Seeflugstützpunkt Rogoznica*; 85x4½ mm; violet to black.

Spalato (now Split, Croatia). In May 1915 a seaplane base was established at the Ivanko shipyards in Spalato. It was referred to as *Seeflugstation Ivanko* and operated under the direction of the station at Sebenico. In 1917, a major seaplane base was planned for Sette Castelli near Spalato. It was to be responsible for the entire central Adriatic coast. But, the Castelli seaplane station did not become operational before the war's end. No postal markings are recorded for the seaplane facilities in the Spalato area.

Curzola (an island; now Korcula, Croatia). In February 1915 a seaplane base was established at the north end of the island at Vallegrande, because of its favorable location on the route toward Cattaro. In June of 1916 it was moved to the bay adjacent to the city of Curzola. It was under the direction of the station at Sebenico. One postal marking of Curzola's seaplane base is recorded.

Curzola Seaplane Station #1 -- One-line; *Kuk Seeflugstützpunkt Curzola*; 79x4½ mm; violet or black. (See Fig. 3-26)

Lagosta (an island; now Lastovo, Croatia). This small base became operational in November 1915 with the mission of neutralizing reconnaissance aircraft flying out of the Italian–held Peligosa islands. It was under the direction of the base at Sebenico. No postal markings of the base at Lagosta are known.

Gravosa (also known as Ragusa; now Dubrovnik, Croatia). Gravosa was for centuries an important port and city. A railroad penetrated the mountains from Sarajevo and Mostar to Metkovich, and then ran southward along the coast through Gravosa to the Bay of Cattaro. This was the only coastal railway along the Adriatic south of Fiume!

The seaplane base at Gravosa was established in August 1917 under the direction of the base at Kumbor. Its mission was mine detection and escorting convoys in the sea lanes from Curzola to the Bay of Cattaro. There are no postal markings reported for Gravosa; but see the note "b" under Kumbor for a piece of mail that possibly originated in Gravosa.

Kumbor (a strait in the Bay of Cattaro; now Kotor, Yugoslavia). The importance of the naval facilities at the Bay of Cattaro has been emphasized over and over again. A temporary seaplane base was established at Teodo on the Bay in 1913, in response to the Albanian crisis. During the summer of 1914, this was hastily converted into a permanent facility known as *Seeflugstation Kumbor*.

This major facility had the mission of protecting the harbor itself and all the approaches to it. It also was responsible for long–range reconnaissance and bombing missions along the entire southern Adriatic coast of Italy to the Cape of Otranto and at the important Allied–controlled port of Valona (now Vlorë, Albania).

Clement lists the markings of Kumbor that read *Flottenflugabteilung* (Float Plane Detachment) separately from those reading *Seeflugstation*. Sobetzky combines them. I've chosen to follow Sobetzky's lead.

Figure 3–34 Seaplane Base Marking of Kumbor

Kumbor Seaplane Station #1 -- Single-circle with eagle; *Kuk Flottenflugabteilung*; 24 mm diameter; red.

Kumbor Seaplane Station #2 -- Double-circle with dual arms; *Kuk Flotten-flugabteilung Kumbor*, 31 mm diameter; violet to blue-black. (See Fig. 3-34)

Kumbor Seaplane Station #3 -- Double-circle with dual arms; *Kuk Flotten-flugabteilung / Kumbor*, 34½ mm diameter; red or violet.

Kumbor Seaplane Station #4 -- Single-circle with dual arms; *Kuk Seeflugstation Kumbor*, 30 mm diameter; red, blue, or black. (See note a)

Kumbor Seaplane Station #5 -- One-line; *Kuk Seeflugstation Kumbor*, 60x3 mm; violet.

Kumbor Seaplane Station #6 -- One-line; *Kuk Seeflugstation Kumbor*, 48x4½ mm; violet.

Kumbor Seaplane Station #7 -- One-line; *Kuk Seeflugstation*; 35x4½ mm; red or violet to blue-black. (See Fig. 3-35 and note b)

Kumbor Seaplane Station #8 -- One-line; *Kuk Seeflugstütz-punkt*; 55x3 mm; violet*. (See note c and Fig. 3-26)

Kumbor Seaplane Station #9 -- One-line; *Kuk Seeflugstation, Etappenpost 267*; 73x3½ mm; dark violet. (See note d)

Kumbor Seaplane Station #10 -- Two-line; *Zensuriert / Kuk Seeflugstation, Etappenpost 267*; 53x10 mm; dark violet.

Notes: (a) Sobetzky lists a variety of #4 with the letters "*ion Kum*" missing. I expect this is only the result of a weak strike.

(b) The card shown in Figure 3–35 is dated July 1918, and bears the exact markings described by Sobetzky for Kumbor. But, it is a picture postcard of Gravosa. That may mean nothing; but again, when did you last send home a picture postcard from a place you haven't been? A logical conclusion might be that it was flown from Gravosa to Kumbor, where it was put into the regular mails. I'll suggest another possibility ... perhaps, the "Kumbor" markings, like #7, that bear no town name are intended for use at other bases under the command of Kumbor -- like Gravosa. Just another example of the study still needed in the postal history of the Seaplane Corps.

(c) The #8 marking is on the courier mail cover shown in Figure 2–26, but is too weak to see.

Figure 3–35 Unusual Card with Kumbor Markings

(d) EPA 267 was an Army FPO that operated with the 19th Corps, at various places along the coast of the Adriatic in 1918. #9 and #10 are the second tie–in between the army fieldpost system and the Seaplane Corps.

Durazzo (also known as Durz; now Durrës, Albania). The seaplane base at Durazzo was established in March 1916, after the capture of the city by the Central Powers. It represented an important strategic location just north of the straits of Otranto. Its mission was to support fleet operations in the south, and to conduct bombing missions in southern Italy.

Durazzo Seaplane Station #1 -- Double–circle with dual arms; *Kuk Flugstützpunkt Durz*; 38 mm diameter; red, blue–violet, or black. (See Fig. 3–26)

Durazzo Seaplane Station #2 -- Single–circle with dual arms (?); *Kuk Seeflugstation Durz*; 24 mm diameter; red.

Durazzo Seaplane Station #3 -- One–line; *Kuk Seeflugstützpunkt Durazzo*; 89x5 mm; black*. (See Fig. 3–36)

Durazzo Seaplane Station #4 -- One–line; *Kuk Seeflugstützpunkt Durz*; 72x4 mm; black violet.

Durazzo Seaplane Station #5 -- One–line; *Kuk Flugstützpunkt Durz*; 67½x4½ mm; black–violet.

So ends the story of Navy Seaplane Corps mail, an area of Hungarian aerophilately that is still very ripe for research and new "finds."

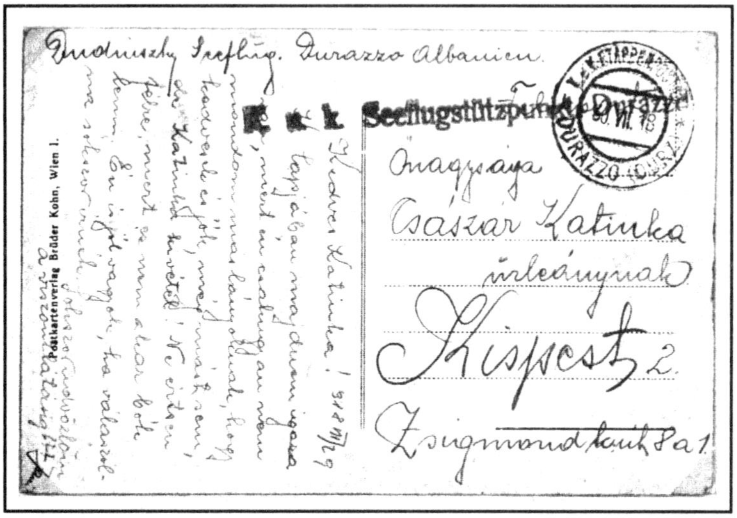

Figure 3–36 Previously Unrecorded Durazzo Marking

Army Aviation and Balloon Units

This section is divided into three parts: the postal markings of Austro–Hungarian Army aviation units; balloon units; and weather stations. These were all under the overall direction of the Army Air Corps, though in the field they were usually "attached" to specific ground military units.

The section on *The Beginnings of Airmail in Hungary* provides background information on the pre–war origins of all these activities. The two major sources of flown mail in World War I -- the *Przemysl Airmail of 1914–15* and the *Military Airmail Service of 1918* -- have their own chapters. So this chapter basically picks up the loose ends.

This chapter primarily concerns itself with the "unit markings" of aviation units on official and personal mail. Virtually none of this mail is flown (with two exceptions I will note shortly). The question of whether most of this material is appropriate for an airmail collection is a personal choice. I find it important because in my collecting of *The Pioneer Period of Hungarian Airmail,* I strove to "document the aeronautical *and* aerophilatelic history of Hungary." The choice is yours.

The two exceptions, noted above, regarding flown mail are: (1) the *very few* pieces of official mail that were forwarded to their destination by courier flights of the units involved; and (2) the *very rare* examples of mail flown by unmanned weather research balloons. These are discussed further in the sections that follow.

The main reason that personal mail was not flown is simply that it was not a part of the mission of the aviation units. Also, since it did not serve any important purpose, there was no incentive for pilots to carry mail as "favors." It wouldn't serve an important purpose because everything the army needed in World War I was carried by railroads, horses, or the troops themselves. As a result, the railroads were well maintained to bring supplies to the lines. There wasn't a lot to bring out except the wounded and the mail. Mail with receiving marks shows that the fieldpost mail system was often quite comparable to the civilian posts in timeliness of delivery. Except for local delivery and military dispatches, the concept of "overnight mail" was for the distant future!

The information provided below is based on the work of Clement (1950's) and Tranmer (1970's), with a few additions from my own research. This area is better researched than the Navy's Seaplane Corps, but I believe that there is still some work to be done and discoveries to be made. Until rather recently, these aviation unit markings were quite inexpensive. Now that prices have increased, I expect some large caches will "come out of hiding" and be the fodder for new studies.

Aviation Unit Markings. For the Austro–Hungarian Army airplane units, I will deal with the unit markings in the following order: (1) units associated with the command structure; (2) front–line operational units; (3) reserve operational units; and (4) administrative and support units.

The first group of aviation unit markings I will discuss are from various headquarters units of the Air Corps. There is no point in trying to list these. They are very numerous and the names changed repeatedly as the army was reorganized. That they were a part of the air forces is obvious from the name of the unit. Examples are: *Kuk Armeeober–kommando / Chef des Luftfahr–wesens* or *Staboffizier der Luftfahr–truppen beim Kommando der Kuk Isonzoarmee*. I don't believe there is a systematic way to collect these. Find them, establish a "Hungarian connection" that satisfies your goals, and that's it!

Much of the mail of interest in the category of aviation unit markings is official and personal mail handstamped with the unit marking of a *Fliegerkompagnie*, the basic operational unit of the air corps in the field. Remember the purpose of unit markings. It was essentially the "ID" required to use the free franking privilege. It says nothing about the mail being flown. In fact, if it is personal mail it was certainly *not* flown. If it is official mail there is an extremely high probability that it was *not* flown. As noted above, before representing a piece as flown (or paying a premium price for flown mail) use your common sense and don't kid yourself.

Another thing to remember is that these aviator companies were field units. Most of the war was not fought on Hungarian soil. (Exceptions were a few minor forays by the Russians in the Carpathians early in the war, and the Romanian incursion into Transylvania in 1915–16. See the chapter on the *Zeppelin Base near Temesvár* regarding the latter.) So, not much *Fliegerkompagnie* mail originated in Hungary. Look for Hungarian addressees or Hungarian–language postal markings ... depending on your "Hungarian connection" criteria.

Most *Fliegerkompagnie* markings come in one of two forms: a two–line marking and a single–circle marking with the Habsburg eagle in the center. Many units used *both* types of markings. Figures 3–37 and 3–38 show an example of each type.

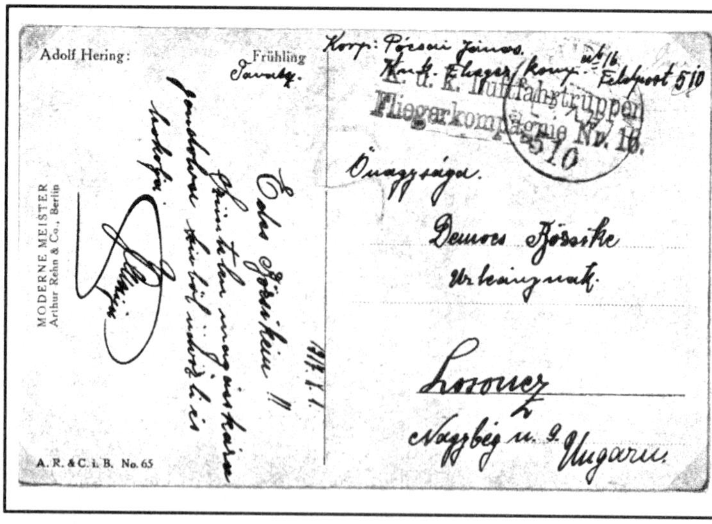

Figure 3–37 Two–line *Fliegerkompagnie* Marking

Figure 3–38 Single–circle *Fliegerkompagnie* Marking

Most of these markings read: *Kuk Luftfahrtruppen / Fliegerkompagnie Nr. XX* where the *XX* is the unit number. Rather than list the numbers that are recorded, I'll mention the exceptions. All unit numbers from 1 to 74 were known to Clement except: 48, 50, 51, and 72. Tranmer claims to have found an example of *Fliegerkompagnie Nr. 48*. In addition, the following numbers are found with a letter appended: 17D, 30J, 55I, 60J, 102G, 103G, 104G, and 105G. Quite probably, the few units that are missing are simply yet to be found.

There are numerous minor variations in *form* of these markings. For example, *Fliegerkom–pagnie Nr. 38* has a one–line marking with the exact same wording; *Fliegerkompagnie Nr. 1* has a two–line marking with the "*Kuk*" on both lines. I have also seen official mail with the unit number simply typed in the return address, rather than being handstamped.

The following variations in wording are reported by both Clement and Tranmer. The marking *Kuk Fliegerregiment / Fliegerkom–pagnie No. XX* is known for the following units: 1, 5, 9, 14, and 17. This marking for *Fliegerkompagnie No. 14* is shown in Figure 3–9 in the chapter on Przemysl. [Note: on this marking, as well as on some others, the word *nummer* is abbreviated *No* instead of the normal *Nr*.] The simple marking *Kuk Flieger–kompagnie Nr. XX* is known for units 1, 2, 4, 7, 11, 12, 14, 15, 25, 35, and 46. The final marking in this group is *Kuk Feldflieger–kompagnie Nr. 8*.

The air corps reserve bases were at fixed locations. Therefore, a "Hungarian connection" for all mail from certain of them is obvious. Below is a list of the markings for *Kuk Luftfahr–ersatztruppen / Fliegerersatz–kompagnie Nr. XX* for all the reserve bases in Hungarian towns.

Reserve Air Corps Units in Hungary	
Nr. 1	Újvidék
Nr. 4	Szombathely
Nr. 5	Szeged
Nr. 7	Pándorfalu
Nr. 9	Arad
Nr. 13	Lugos
Nr. 15	Arad
Nr. 16	Eger

The following additional markings are listed by Clement.

Kuk Luftfahrersatztruppen
Kommando des Fliegerersatzbataillons No. II

in Arad and No. IV in Újvidék. See Figure 3–39 for an example of this Arad marking. Also Clement lists:

Kuk Luftfahrerstaztruppen
Fliegerersatzbataillons No. II
Fliegerersatzkompagnie 13

in Lugos. I have found a similar 3–line marking, previously unrecorded, of:

Kuk Luftfahrersatztruppen
Fliegerersatzbataillon
Fliegerersatzkompagnie Nr. 1

in Újvidék. It is illustrated in Figure 3–40. I'm sure there are many others yet to be recorded.

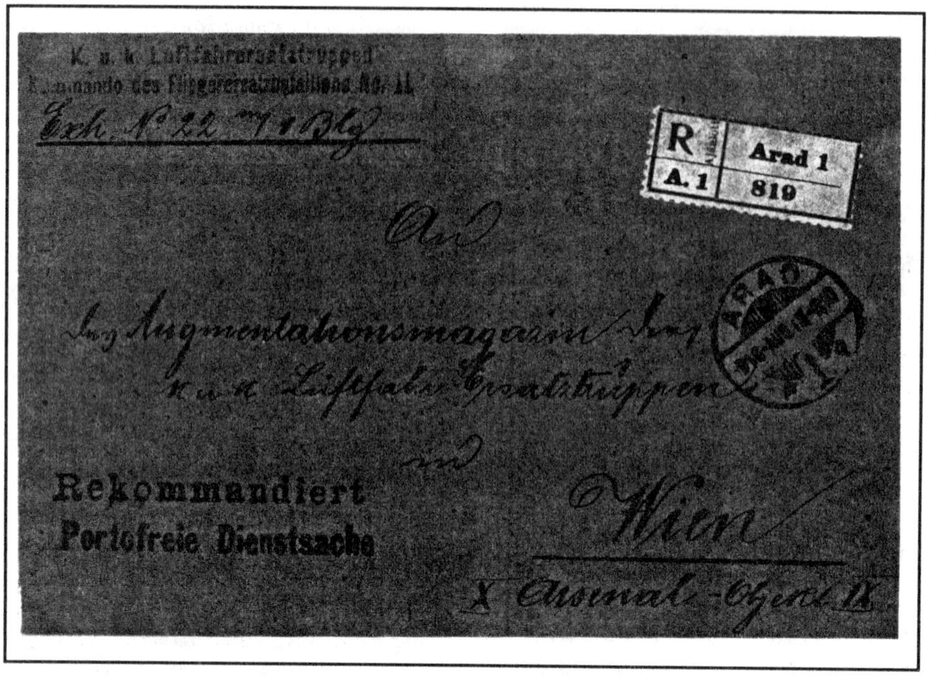

Figure 3-39 Marking of the Reserve Battalion at Arad

The various administrative and support functions are more difficult to categorize. One group is the *Fliegerpark*'s -- aviation depots. Clement lists several markings, but does not identify their locations. The following markings are noted: *Kuk Luftfahrtruppen / Fliegerpark Nr. 1*, *Nr. 2*, *Nr. 6* and *Nr. 7*. Also, *Kuk Luftfahrtruppen / Fliegeretappenpark Nr. 1* through *Nr. 9*. *Fliegeretappenpark Nr. 2* was in Szeged. The others I'm not certain of. There were also depots named after notables. Figure 3-41 shows a card with the unit marking of *Kuk Fliegerpark / des HGK. FM. Frh. v. Conrad*, named for the Army Chief of Staff, Field Marshal Ferenc Conrad. Also, the previously unrecorded marking *Kuk Fliegerpark / des HGK. GO. Erzh. Josef*, named for the Archduke Joseph, exists. There are certainly other markings associated with a variety of aviation support units. Support units are also listed with the balloon unit markings. I believe some of the support facilities had roles related both to airplane and balloon units. So, organizing your research, as well as your material, is not an easy task.

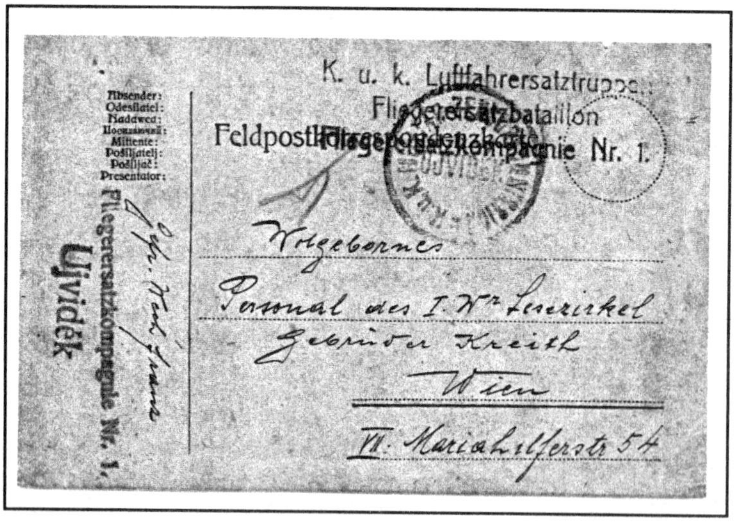

Figure 3-40 Marking of Reserve Company at Újvidék

There are other unit markings that are related to aviation in only the most tenuous way. Among them are the markings of certain air defense units. An example is the unit marking for the *Kuk Luftfahrzeug Abwehrkanonen / Zug No. 8* ... an anti-aircraft gun unit. These are neither aviation units, nor operationally a part of the air corps. But, the bottom line is always "collect what you want to collect."

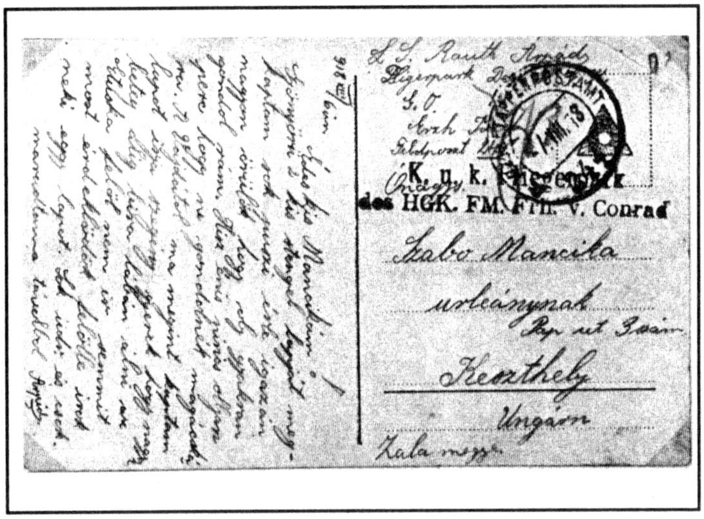

Figure 3–41 FM Conrad Unit Marking

Balloon Unit Markings. The dirigibles operated by the Austro–Hungarian Army were obsolete when the war began. The Zeppelins ordered for Austria–Hungary had not been, and never would be, delivered. Therefore, the *Luftschiffer* units were strictly balloon units. As previously noted, their principal mission was artillery observation, with a limited role in reconnaissance and coastal defense.

The likelihood of mail with balloon unit markings being flown is even lower than for aviation unit markings ... essentially nil. If anything was flown, it was certainly only official mail that was being expedited to headquarters. For example, in my collection is an official cover of balloon detachment 5 that was posted near the front in Russia and was delivered at headquarters in Debrecen the next day. It moved fast to be sure. "Possibly flown" is the best you can say!

As with the heavier–than–air side of the air corps, there are many unit markings associated with the command structure. An example is: *Kommando der Kuk Isonzo–Armee / Ballonstabsoffizier*. Again, there were too many such headquarters units, and the names changed too frequently to even attempt to enumerate them.

The basic operational balloon unit was the *Ballonkompagnie* or the *Ballon–abteilung* (Balloon Company or Balloon Detachment). The markings tended to be either the standard two–line marking or an oval marking with the Habsburg eagle in the center. Again, many variations of both form and wording exist. Figure 3–42 shows an example of the two–line marking. Balloon Company 27 was a coastal defense unit at the Bay of Catarro. Figure 3–43 shows an example of both a two–line marking and an oval marking for the same unit -- Balloon Detachment 14 that did artillery spotting in Galicia.

The unit numbers reported by Clement are as follows. *Kuk Luftfahrtruppen / Ballon–abteilung Nr. XX* where the unit numbers recorded are 1, 5, 7 through 17, and 20 through 23. *Kuk Luftfahrtruppen / Ballonkompagnie Nr. XX* where the unit numbers are: 1 through 14, 16, 18 through 21, 23, and 25 through 27. Obviously, some are yet to be "discovered."

Another group of markings read *Kuk Luftschifferabteilung / Fliegerkompagnie Nr. XX* where the units recorded are 1–4, 8, 10–12, and 15. To me, these imply that the balloonists had some airplanes at their disposal, perhaps for logistics and supply.

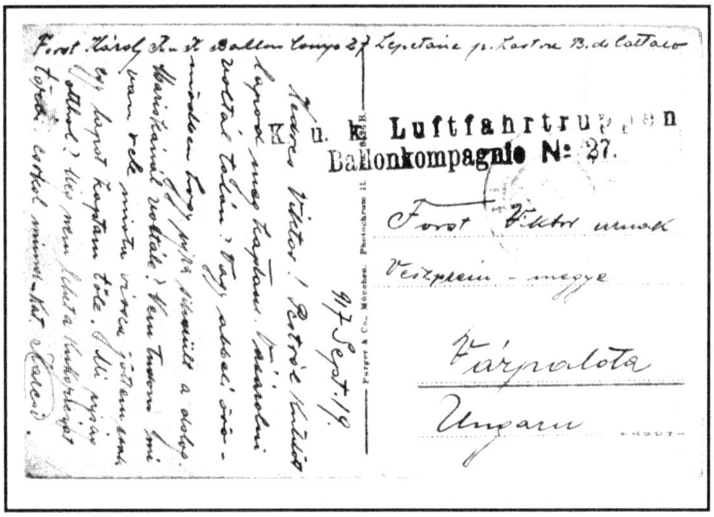

Figure 3–42 Two–line Balloon Unit Marking

Figure 3–43 Two–line and Oval Balloon Unit Markings

Weather stations in the field operated generally with aviation and/or balloon companies. The basic unit marking was a two–liner reading: *Kuk Luftfahrtruppen / Feld–wetterstation Nr. XX.* The unit numbers recorded by Clement and Tranmer are 1–3, 5–16, and 45. Additional discoveries will yet be made. More research is needed. Figure 3–44 shows an example with the markings of *both* Weather Station #1 and Aviator Company #26.

A red double–circle marking reading *Kuk Festungs Ballonabteilung 1/3B* from a harbor defense unit at Pola, and many other variations of balloon unit markings exist ... some of which are reported by Clement or Tranmer. I see no point to repeating these lists here, since they are clearly incomplete.

As with the airplane units, many administrative and support units existed in the lighter–than–air operations of the Air Corps. One example recorded in Tranmer is *Kuk Luftschiffer–abteilung / Flugpark No. 13.* Another support unit with a distinct Hungarian connection is *Kuk Luftschiffer– / abteilung / Flugzeugdepot* which was located in Satoraljaújhely. The name of this unit again indicates that the *Luftschiffer* had some airplanes at their disposal, or that this unit served both parts of the air corps.

Weather Station Markings. The weather service also had its headquarters functions. They were called *Feldwetterzentrale*, and unit markings exist. To the best of my knowledge, none of these were in Hungary.

In addition to weather stations near the front lines, stations were located at military bases. Certainly, one was located at Szombathely. I am uncertain whether a specific unit number was assigned to these, but it's probable the station at Szombathely in 1916 was #6 (see below).

With the war on, the costly manned balloons were needed to support military operations. Most weather research was stopped. An exception are the drop cards of 1917, pictured by Alföldi and Talas, from the balloon

Figure 3–44 Weather Station Unit Marking

Hainburg. These wartime dropcards were postfree. They are very scarce.

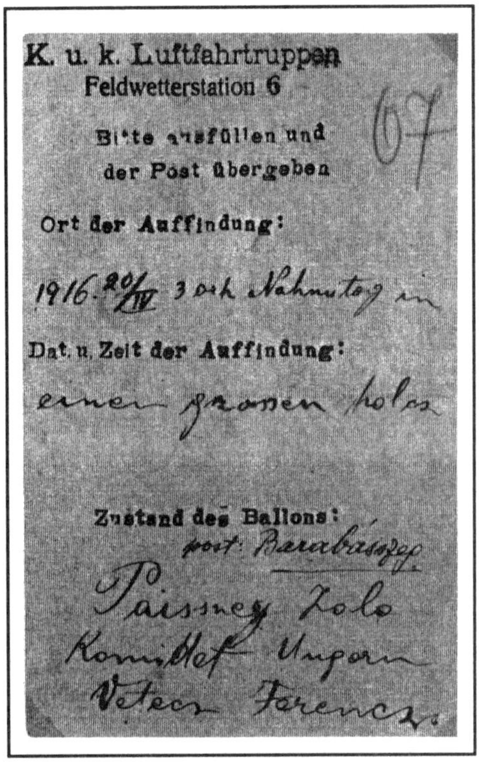

Figure 3–45 Unmanned Balloon Card

The most aerophilatelically interesting aspect of weather station operations was the use of unmanned balloons for continuing weather research. Where these unmanned balloons were released is unknown -- probably Szombathely. They carried a card to be posted by the finder, with information about the location and time they were found. Obviously there would be no purpose to sending more than one card with each balloon, so they are certainly quite rare.

I have seen two types of these flown balloon cards. The first -- of which the only recorded example is pictured in Figure 3–45 -- was posted to *Kuk Feldwetterstation Nr. 6* in April of 1916. The second type, of which I've seen three examples all dated in late 1917 or early 1918, was addressed to the *Kuk Feldwetterstation in Szombathely*. Figure 3–46 shows one of these.

The first of these types is a salmon–colored Fieldpost card. On the front it is simply addressed to *Kuk Feldwetterstation No. 6* using a 70 mm one–line, black handstamp. "Feldpost No 606" is typed below that. It also bears a rectangular postmark of Barabásseg of April 21, 1916. This village is south of Zalaegerszeg in Zala megye.

The reverse of the card, shown in Figure 3–45, has the normal 2–line unit marking of Weather Station #6 at the top, with typed questions below. The finder provided only minimal answers to these questions before posting the card. From this single example, it is impossible to tell much about how these cards were used.

The second type provides more information. It is printed on pale green card stock with dimensions of 140x185 mm. This is folded in half to become postcard size. Both the front and back are printed, rather crudely, probably by mimeograph or a similar process. The reverse, which is not shown, gives directions to the finder in three languages: German, Hungarian, and Croatian. Roughly translated, those directions are:

> The finder is requested to fill in card and mail it postfree – the balloon will not be reused.
>
> 1) Place found: nearest village or town and county and if possible the longitude and latitude and altitude above sea–level.
>
> 2) Time balloon was found if it had been on the ground for some time.
>
> 3) Time balloon was found if it was observed.
>
> 4) The address of the finder.

Figure 3–46 shows the front of this type card unfolded. Note that the printed address is to the *Feldwetterstation in ....* -- leaving the town name to be filled in. This indicates that they were prepared for use in more than one place. Therefore, like the balloon cards of 1908–12, they appear to be a form of official postal stationary. But, all recorded examples have Szombathely written in. In addition to the town name, the balloon number was written in prior to lift–off.

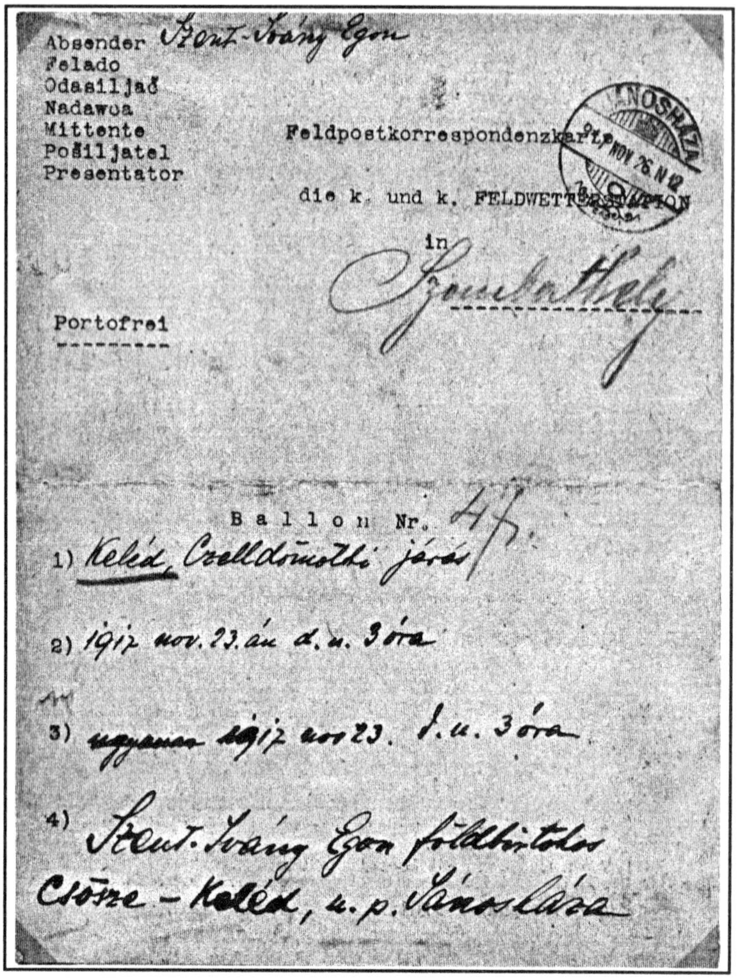

Figure 3–46 Folded Card Flown by Unmanned Balloon

The card shown was found at Keléd and was posted on November 26, 1917 at Jánosháza. Both villages are south of Celldömölk in Vas megye.

Another example I'm familiar with was posted at Alsórámócz in Sopron megye in March 1918. This is now the town of Unter Rabnitz, Austria. The finder was extra diligent. He even provided a little map of where he found the balloon in Répczebánya (now Piringsdorf).

Since all known copies were posted in the western part of Hungary, it is generally thought that the weather station at Szombathely created these cards and exclusively used them on unmanned balloons.

Zeppelin Base near Temesvár

During World War I, German units operated in Hungary. They were particularly prevalent in the Carpathian mountain campaigns of 1915 and 1916. Here they attempted to assist Austria–Hungary in stopping the Russian and Romanian incursions from the north and east.

A particularly important part of this effort was the attempt to repel the Romanian advance into Transyl–vania. This campaign was most intensive in the Winter of 1915–16. It was over by September 1916 when the Romanians were driven out of Hungary and put on the defensive on their own soil.

In 1915, the Romanian advance had reached its deepest point. Three major cities: Brassó, Nagyszeben, and Marosvásárhely were in their hands. Temesvár and Kolosvár were threatened.

The Zeppelin LZ–81, and other Zeppelins, were sent by the Germans to Temesvár to protect the city from Romanian air raids, and to bring the war home to Romania by bombing its cities. A photo, from a picture postcard, of the LZ–81 docked in Hungary is shown in Figure 3–47.

A Zeppelin Base was established on the Hungarian plain outside Temesvár, near the village of Szent András. According to Sieger,

Figure 3–47 The LZ–81 at Szent András

the following Zeppelins flew from that base during 1915 and 1916 while it was operational: the LZ–46, LZ–81, LZ–85, and LZ–97. I will briefly describe the activities and fate of each of these Zeppelins.

The LZ–46 was staffed by a detachment from the *Königl. Sächsen Feldtruppen* (Royal Saxon Expeditionary Forces). This dirigible, also known as *Luftschiffe Nr. 14*, was lifted by 31.9 km^3 of gas. It was launched August 9, 1915. After being employed for reconnaissance and attack missions from Hungary, it survived the war as a training airship for the German Navy.

Figure 3–49 German Szent András Cancellation

The LZ–81 was manned by a detachment from the *Kgl. Württ. Luftschiffkommando* (Royal Württemberg Airship Command). It was commissioned on October 7, 1915 and was found to be underpowered. It was rebuilt with the gas bags expanded to 35.8 km^3 for its 1916 bombing raids against Romania. It crashed at Tetevo, Bulgaria (about 40 miles south of Bucharest) in September 1916, after suffering anti–aircraft gun damage while attacking Bucharest.

The LZ–85 was a 31.9 km^3 airship launched on September 12, 1915. In May of 1916, while based at Szent András, it was brought down during a raid on Salonika, Greece and fell into British hands. Figure 3–48 shows a card from a crew member of the LZ–85, posted at the Szent András base.

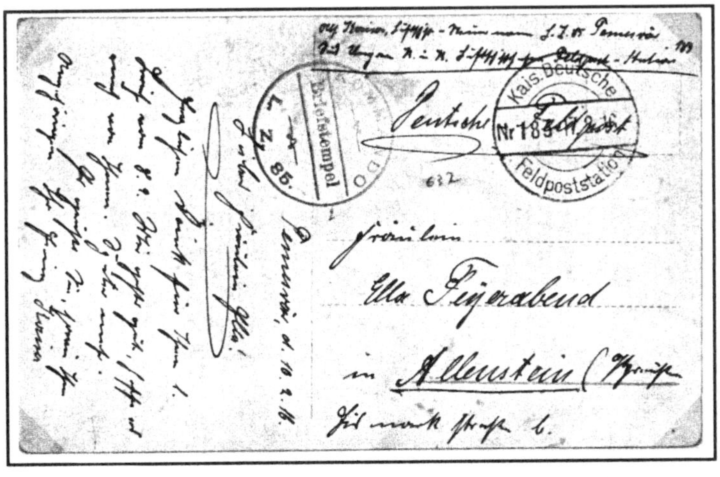

Figure 3–48 LZ–85 Mail from Szent András

The LZ–97 was a 35.8 km^3 Zeppelin launched on April 4, 1916. After performing attack missions against England and on the western front, it was transferred to Szent András where it conducted raids against Romania. It was decommissioned in July 1917.

During the period when the Zeppelin Base at Szent András operated, German troops used *Kais. Deutsche Feldpoststation Nr. 183*. Austro–Hungarian support troops used the civilian postal system, where their free–frank privileges still applied. I have in my collection mail posted at Szent András by crew members of the LZ–46, LZ–81, and LZ–85 -- each with the Zeppelin's unit marking and a postmark of the German fieldpost office #183. For the LZ–97, I have a card with a manuscript notation of the sender's airship and a postmark of German fieldpost office #16. It was mailed in Budapest on March 14, 1917. Apparently the LZ–97 was in transit from Temesvár back to Germany at that time, or was on a temporary assignment in Budapest.

Figure 3–49 shows another interesting German piece from Szent András that is in my collection. It has a marking of *Luftschiffertruppe 14* at the *Luftschiffhafen Temesvár Südungarn* (Airship Detachment 14 -- presumably the LZ–46 -- at the Temesvár Airship Base in South Hungary).

This postcard to Berlin is franked with Hungarian stamps that are postmarked with a German double-circle cancellation that reads *Luftschiffhafen Szent András bei Temesvár / Sudung.* (Airship Base at Szent András near Temesvár.) While undated, this card sends "New Year Greetings" for 1916. The used "Soldiers' Christmas" label was added before mailing, since it is tied by the cancellation.

As noted above, Austria–Hungary provided support troops to assist at this German Zeppelin Base. Their fieldpost cards show the markings of one of two units: the *Kuk Luftschiffertruppe / Szent András* or *Kuk Kommando "Adebar"*. They are usually postmarked with the normal civilian postmark of Temesvár. No stamps were required for fieldpost mail within the Austro–Hungarian Empire. These markings are quite unusual.

The first of these units was a logistics and support unit assisting the German airmen. One of its biggest tasks was hauling down and docking the Zeppelins. Considering the size of these airships, this was a formidable and dangerous task. Two unit markings are known. They are the circular marking shown in Figure 3–50 and a two–line marking also reading *Kuk Luftschiffer–Truppe / Szent–András*.

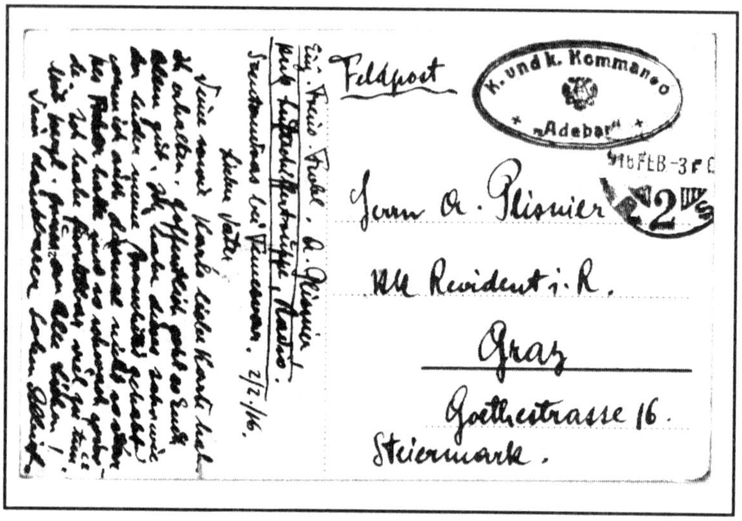

Figure 3–51 Marking of Radio Station at Szent András

The second of the unit markings is rather unusual. It is the marking of the Austro–Hungarian operated radio station at the Zeppelin Base. Its "call name" was *Adebar*. One of the three known markings from this unit is the oval handstamp shown in Figure 3–51. The message on this card first identified to me that *Adebar* was a radio station. Additionally, there is a blue two–line marking *Kuk Adebar–Meteor / Temesvár* and a one–line marking *K. und k. Kommando "Adebar"*.

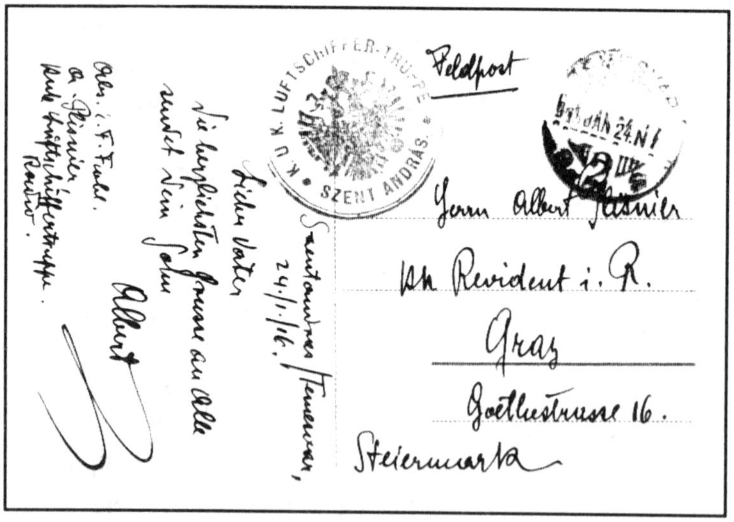

Figure 3–50 Airship Support Unit at Szent András

The markings of the Zeppelin Base at Szent András were not previously known to most researchers in Hungarian aerophilately. Though they are not *flown* airmail, they represent an extremely important aspect of the aeronautical history of World War I in Hungary. I expect that I have just brushed the surface of this subject to date -- and, it is a subject of interest to a much broader community than simply Hungarian airmail collectors.

Army Airplane Exhibition of 1917

There are areas of Hungarian aerophilately that are neither esoteric nor expensive. One of these "fun" areas is the "mail" of Budapest's 1917 *Hadi Repülö Kiállitás* -- Army Airplane Exhibition. While only one item is flown, and it is rare, most of the material gives the collector a good sense of aviation history and is relatively inexpensive.

At the beginning of the war, Austria–Hungary's air forces were in a very poor state. As already noted, they developed quickly primarily based on the importation of German aircraft designs. These were licensed for manufacture in Austria–Hungary itself. By 1917, there was a large, powerful, and profitable aviation industry that had operations in both the Austrian and Hungarian portions of the Empire. Despite the war, these companies were competing against one another for the most lucrative contracts. The airplane exhibition of 1917 in Budapest, which some claim produced Hungary's "first airmail," provided these companies a showcase for their products.

The exhibition was held at the Budapest City Park's Hall of Industry. This is one of the buildings constructed for the 1896 Millennium celebration. It was a large open building with a high ceiling, and was lighted by windows around the entire roof line. It could easily accommodate displays containing many of the relatively small airplanes of the period. Figure 3–56 provides a partial view of the interior of the hall.

The opening date of the exhibit was April 29, and it was originally scheduled to last a month. The exhibition actually ran a full two months, closing on July 1. A postoffice counter was provided at the exhibit during the entire period. Three items were prepared for use on mail posted at the exhibition site: a special cancellation, a registration etiquette, and a vignette.

The vignette was privately printed by "GLOBUS" of Budapest, probably for the management of the exhibit. They were printed on white paper and perforated 11½. The sheet margins were usually imperforate, so vignettes that are imperforate on one or two sides are common. Fully perforated, and fully imperforate sheets are also known.

The 30x45 mm vignette design is black, blue and orange. It pictures an airplane and a Zeppelin silhouetted in the searchlights of a ship. The vignette is found on most -- but not all -- of the mail canceled at the exhibit. A publicity flyer, which measures 168x119 mm was printed in the same design and colors. The vignette can be seen in Figure 3–52 and 3–57 below.

The "mail" canceled at the exhibit hall was postmarked by a Budapest circular date stamp, with *Hadi Repülö Kiállitás* at the bottom. It was primarily used on canceled–to–order items that were handed back over the counter at the exhibit. The postmark usually canceled stamps, but is sometimes also found tying vignettes, charity labels, etc. to a piece of mail.

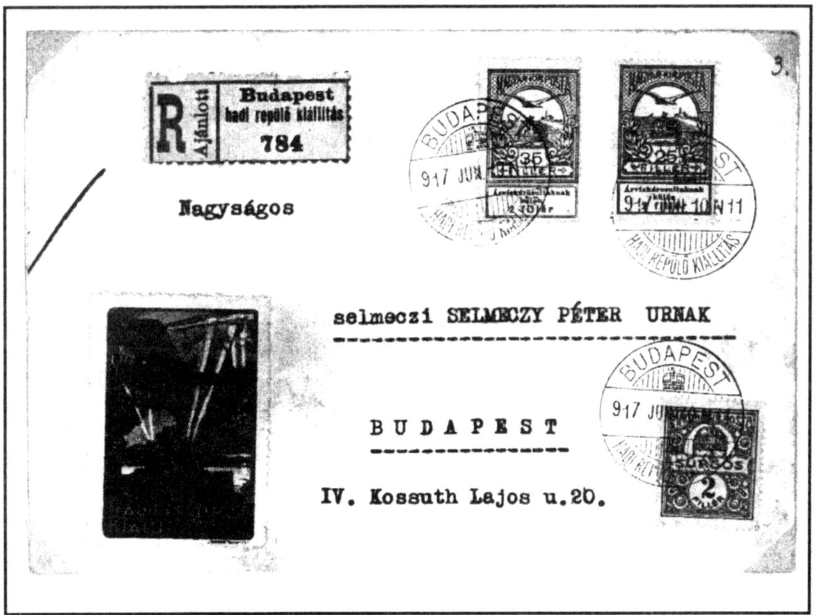

Figure 3–52 Airplane Exhibition Mail

The final philatelic item of the exhibit is a registry etiquette. It is of the standard sort, but is inscribed *Budapest / hadi repülö kiállitás*. Since little actual mail was posted from the exhibit, the etiquette is relatively scarce. Figure 3–52 shows all three of these philatelic items on a local registered cover.

To many collectors, the only aspect of Airplane Exhibition philately they have seen is the stamp -- probably a 5 fillér Harvester -- with a "rare cancellation" tied to "a piece." It is noteworthy that these did not come from envelopes, as one normally expects for "stamps on piece." Actually they are cut-outs from a sheet of company stationary that was covered with stamps and canceled-to-order at the exhibit's postoffice counter. Figure 3–52 shows an example before being cut up and put on sale, to less-than-knowledgeable collectors.

Of special interest are first and last day cancels of the exhibit. First day cancels -- April 29 -- are found on cards prepared by Pál Rampacher (a well-known philatelist and later a jurist). These were intended for a flight from Budapest to Miskolc to be flown that day, a flight that apparently never took place. As first day usages, they remain of special interest. An example is shown in Figure 3–54. A manuscript message on the reverse of this card translates to: "First exhibit mail for Miskolc special flight."

Figure 3–53 The Origin of Exhibit "Stamps on Piece"

On the last day of the exhibit -- Sunday, July 1 -- a short mail-carrying flight was made over Budapest. It landed in the nearby village of Pilotaújfalu. Some people consider this Hungary's first airmail. I don't subscribe to that theory. Obviously, such characterizations depend upon one's definitions. The flight was by an airplane (rather than a balloon) and it was completely within Hungary (unlike the 1914 Przemysl–Kassa flight). These are useful distinctions.

The two pieces of mail, that I have seen from this flight were both canceled at the exhibit on June 30. They bear the receiving mark of Pilotaújfalu of July 1. July 1 mail may also have been carried. Figure 3–55 shows a cover carried on the Budapest–Palotaújfalu flight. The other cover I've seen was not addressed to Pilotaújfalu. The amount of mail carried is unknown, but it is quite scarce. Other mail with last-day (July 1) cancels may be of interest to some collectors.

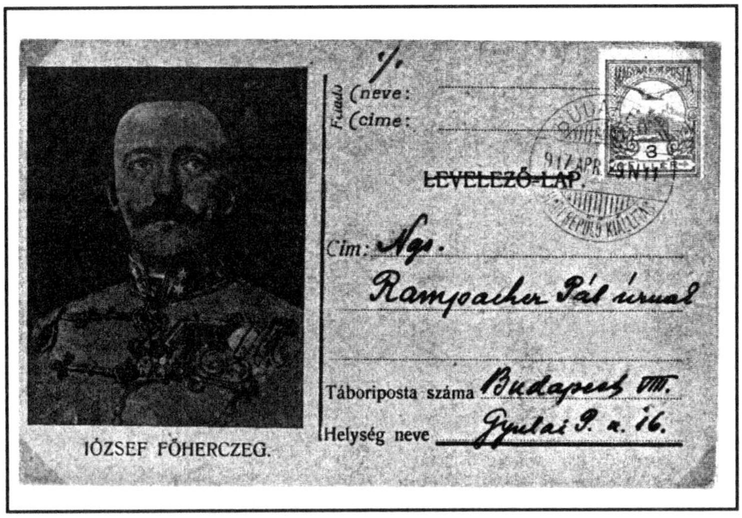

Figure 3–54 Exhibition First Day Cancel

A most interesting aspect of this exhibition was its use for advertising by military contractors. President Eisenhower coined the phrase "military–industrial complex" after the Second World War. The philatelic material of the 1917 Budapest Army Airplane Exhibition proves that a very similar establishment not only existed, but was thriving, in the Austro–Hungarian Empire forty years earlier.

Figure 3–55 Budapest–Pilotaújfalu

Most of the "mail" canceled at the exhibit consisted of postcards with stamps affixed, canceled–to–order. Many of these postcards fail into two categories: (a) the advertising cards of aviation contractors; and (b) a historical aviation series of cards picturing aspects of the 1910 flying competition. Certainly, postcards were canceled that fall into neither of these categories. I will mention a few of these at the end of the chapter, since most of them are also of interest to the aviation enthusiast.

The postcards of military contractors go from the obvious (e.g. a picture of the contractor's exhibit), to the subtle (e.g. a picture of the head of the company with the King), to the exaggerated (e.g. a card with the inscription "Destruction of the Arsenal at Venice by our bravest Navy pilot, using *Hiero* airplane engines;" not planes, not bombs, just engines!)

These cards are shown on the next page. Figure 3–56 shows the MAG exhibit in the hall. The inscription translated reads, "The *Hungarian General Machine Company* (Budapest–Mátyásföld) area at the Army Airplane Exhibit." Figure 3–57 shows both sides of a card picturing "Our King conversing with retired Lt. Hendrik Bier, veteran pilot and general manager of the *Hungarian Lloyd Aircraft Factory*." The front is typical of the CTO cards, but happens to be canceled on the last day. Keep in mind, just because someone wrote a name and address on such a card (perhaps decades after 1917) doesn't mean it went through the mails. Figure 3–58 shows the card picturing the aftermath of the bombing of Venice that was noted above.

To date no one has attempted to enumerate these aviation contractor advertising cards. It would be a "fun" project. I'm certain there are many cards that I haven't seen yet, residing in collections around the world.

The second major group of 1917 Army Airplane Exhibition postcards picture historical events associated with the 1910 International Flying Competition.

Figure 3–56 MAG Exhibit

Figure 3–58 Bombing of Venice

Figure 3–57 Lloyd Airplane Factory Advertising Card

In each case in this aviation historical series, the description of the picture is at the upper left on the front of the card. At the lower left is a statement to the effect that the price of the card will benefit the national aviation assets. Figure 3-59 shows an example of one of these. It is inscribed: "The hangar village at Rákos."

Figure 3-59 Hangar Village at Rákos

Finally, I will mention and try to categorize a few other cards from the 1917 Army Airplane Exhibit that I've seen. I will not picture them, simply because this group of cards represents a very open-ended list.

One card I've seen shows a part of the exhibit that apparently was more patriotic than commercial. It pictures a captured French seaplane and mentions that its pilot was the "renegade son of an Istrian medical doctor."

Another shows the airplane of Antal Lányi in flight. The vintage of the photo is clearly pre-war, but it is not part of the historical series discussed above. Yet another shows an aerial photo and is labeled "Aerial observation of the fire effects of bombs dropped on Russian heavy artillery." While showing military action, it doesn't mention any specific contractor.

Alföldi and Talas attempt to list all of the postcards in their catalog. I believe such a list is futile, because many readers will assume it is definitive and complete. It is not. I'm sure many additional items will be recorded in the future. Happy hunting!

Military Airmail Service of 1918

Many people consider the daily airmail service between Budapest and Vienna in July 1918 to be the first real international airmail service. As you already know from reading this far, I am hesitant to state absolute conclusions. But, on this one I tend to agree. Army air courier services -- for military dispatches, etc. -- had been operational between Vienna and Budapest throughout the war. The difference in July 1918 was that the Army pilots were to fly civilian airmail on a regular daily schedule. Let's go back to the beginning.

Background. I feel the best way to introduce you to this very significant service is to provide information from the "horse's mouth." Capt. August Raft-Marwill organized this airmail service and was aboard the first flight of July 4, 1918. Below, I give you translated quotations from a 1967 personal letter from Raft-Marwill to the late John J. Latham, BEM, who was a good friend and a well known British aerophilatelist.

I also tell more of the story by commenting on each quotation. Raft-Marwill's words are printed in italics, so you can easily differentiate between statements from his letter and my comments. Remember, though, these were his informal recollections fifty years after the fact. After excerpts from the letter, I will move to philatelic examples from the 1918 airmail service.

[Note: I have seen Capt. Raft-Marwill's name spelled various ways. In his letter to Latham, I swear his signature is "Raft-Marvil.". I now possess a letter to him, from his mother. In her handwriting, it is obvious they spelled the name, at least during the war, "Raft-Marwill." She was the wife of the Austro-Hungarian General Marwill who commanded the Army base at Zalaegerszeg. She used the name Marie von Raft. Capt. Raft–Marwill's unusual name, therefore, is a precursor to the modern custom of hyphenated surnames. Figure 3–60 shows a photo of Raft–Marwill in 1918.]

"In the Spring of 1918, the airmail line from Vienna to Kiev was organized on orders from the Austro-Hungarian War Ministry. As organizer and director of that airmail service, I (then Captain in the 4th Regiment Horse Artillery and a Pilot in the Airborne Troops) was appointed, and in April of 1918 I was already busy preparing for an additional airmail line from Vienna to Budapest [continuing to] Odessa."

Figure 3–60 Capt. August Raft–Marwill

The Ukraine emerged in 1917 from the Treaty of Brest–Litovsk, as an occupied, but "independent" state. To maintain regular contact with their occupying forces, an airmail service from Vienna to Kiev via Cracow, Lemberg, and Proskurow, was established. Very shortly the Austrian portion of this route (Vienna–Lemberg) was opened to civilian mail. Austria issued airmail stamps; and mail between Vienna, Cracow, and Lemberg was carried in significant quantities for over six months.

Time–Life's **Epic of Flight** (certainly not an authoritative source) tells unbecoming stories about Raft–Marwill. According to them, when he made the initial 3–day survey flight from Vienna to Kiev, he was amazed at the quantity and quality of the food offered at his hotel. He made the return flight in one day, an unheard of feat in that era, to preserve the freshness of the 11 pounds of caviar that returned with him. The same article notes that he grumbled at the additional demands placed upon his men and aircraft when the Hungarians insisted upon having their own airmail service.

The second route was intended to provide speedy contact with the major Ukrainian Black Sea port of Odessa, via Hungary and occupied Romania. At least in part, it was justified by the demands for *equality* by the Hungarian portion of the Dual Monarchy, which had become increasingly influential during the years preceding World War I. Also, throughout the war, a military courier service had operated between Vienna and Budapest carrying dispatches. The second airmail line had both military and political significance, and did not seem to be particularly difficult to implement.

"The first segment of this planned airmail line, from Budapest to Vienna, was opened on 4 July 1918. I piloted the first flight from Budapest (Mátyásföld airport) to Vienna (Aspern airport), and my companion was Lieutenant and Pilot Emil Varga. We flew a 2-seater Brandenburg type biplane, with a 200hp Hiero aeroengine. Our flight lasted from 10.30 to 12.33 hours, arriving 20 minutes late because we had to fly around a storm."

Only the Vienna–Budapest segment of this planned route was ever instituted, and that for only three weeks in July of 1918. The airmail service was available to normal civilian mail as on the Austrian route, but Hungarian registered mail was also carried. This was not the case for Austrian civilian mail.

On the first day, the only flight was westbound. Subsequently there was an eastbound flight leaving Vienna at dawn, and a westbound flight leaving Budapest in the late afternoon. The last

Figure 3–61 Putting the Mail Aboard

flight from Budapest was on July 23. The last day airmail service operated for civilian mail was July 24. The only flight that day was eastbound.

Raft–Marwill claimed to pilot the first flight. Many Hungarians say Varga was the pilot. Unfortunately for Lt. Varga, he did not live to press his claim. He was killed in a crash on the same airline nine days later. Capt. Raft–Marwill lived sixty more years. Also, he outranked Varga, so if he wanted to be pilot, he could be! The real point is that this was an *international* endeavor. That's why both an Austrian and a Hungarian were on the airplane. So, does it matter who held the controls?

The photo in Figure 3–61 shows the first flight mail bags being lifted aboard the aircraft to Lt. Varga (or is it Raft–Marwill?).

"The Hungarian Post Office issued its own airmail stamps with the overprint Repülö Posta (airmail) and 1K50f."

Hungary issued two overprinted airmail stamps, which are listed by Scott as C1–2. Their denominations were 1K50f and 4K50f. These stamps were only valid for paying the airmail surcharge, not other postage fees. The airmail surcharge *had* to be paid with these stamps. Patrons paid the surcharge in cash and the postal clerk put the stamp(s) on the mail.

These stamps were created by overprinting regular values of the *Parliament* issue. The 1K50f value was created by overprinting the blue 75 fillér stamp in carmine. A "brick red" overprint variation is known. The 4K50f (illustrated in Figure 3–62) was the result of overprinting the 2 Korona *Parliament* stamp in dark blue. It was reported that 70,000 of each were printed. Only after the service ceased operations were these stamps sold openly for philatelic purposes. The actual number sold is unknown, but unused examples are worth far less than used stamps on flown covers.

As with all the airmail stamps we will discuss, a few imperforate stamps were sold from the

Figure 3–62 The High Value of the 1918 Airmail Issue

Figure 3-63 Imperf Examples of the 1918 Airmails

postal archives after World War II to acquire hard currency, except for the 1920 issue, which never existed imperf. Reportedly, the number of each sold was 300. Figure 3-63 shows examples of the 1918 imperfs. I will not illustrate, or even mention, imperforate examples of the other airmail issues of the 1920's and 30's.

Postal order 59.622 found on page 284 of the *Official Gazette* for the Hungarian Post and Telegraph Service of July 2, 1918 (No. 63) was the decree that announced the organization and regulations of the airmail service. On July 17, an amendment confirmed the intent of the original decree, that the airmail stamps could not be used for any postal purpose other than the airmail surcharge.

This decree also noted that the postal service reserved the right to transport telegrams by the fastest means available. In other words, telegrams might be carried by airplanes rather than sent over the wires. In fact, they were; but, telegrams carried by airplanes are not identifiable.

The airmail surcharge was 1K50f for each 20 grams, for each flight segment. The maximum airmail surcharge for normal weight mail was therefore 4K50f, for the three segments from Budapest to Vienna and then on to Cracow and Lemberg. Only official and military mail was carried on the Lemberg–Kiev segment.

In addition to the airmail surcharge, normal postage was required (10f for postcards and 20f for normal weight letters). The 60f special delivery fee was required, and the registry fee of 25f was charged when applicable. Finally, there was a 1K charge for the automobile trip from the Budapest main post office to the airport. As an example, a normal weight letter flown from Budapest to Vienna was franked with the 1K50f air mail stamp, plus 1K80f in regular postage (1K auto carriage + 60f special delivery + 20f normal letter rate). Note that the only correct usage on a normal weight letter for the 4K50f stamp is to Lemberg.

Figure 3–64 shows the two pilots in the aircraft, preparing for takeoff, with some of the many dignitaries present that day standing by.

"On the first flight, we carried 520 letters and 300 telegrams. On 13 July and on 21 July, however, there were two crashes on this segment, in which on 13 July the aforementioned Lieut.-Pilot Varga with his companion Lieut. Winger ... were casualties."

Figure 3-64 Preparing for Takeoff -- July 4

The number of pieces carried each day is given in the aerophilatelic portion of this chapter that follows. The mail survived both the crashes mentioned above, though the pilots, and as we shall see, the airline, did not. The crash of July 13 occurred at Mosonmagyaróvár, near the Austrian border. The mail was carried the rest of the trip to Vienna by rail. The crash on July 21 occurred at takeoff in Budapest, killing Sgt. Németh and Lt. Tomaschek. Another plane flew the mail one hour later.

"This and other circumstances led to the conclusion that the Danube monarchy, already at that time near the end of its capacity, was no longer able to meet the demands of this airmail line technically. Therefore, on July 24, 1918 it was abandoned."

Crashes and technical difficulties certainly were factors in terminating the Budapest–Vienna airmail service. But, the service from Vienna to Kiev continued. Apparently "capacity" was the most significant issue. Budapest–Vienna was largely politically justified, and served no critical military need since the line was never extended to Odessa, as originally planned. A military air courier service continued to operate between the capitals of the Dual Monarchy until the last days of the war.

Cities Serviced. The Austrian route was flown daily from Vienna to Cracow and on to Lemberg. Civilian mail was not permitted on the Lemberg–Kiev leg. It operated from late March until October 1918.

The Vienna–Budapest route was flown daily in each direction ... but, only for three weeks in July. Other than for the first flight, the airplanes left Vienna at dawn and arrived in Budapest early in the morning. They then left Budapest late in the afternoon, arriving in Vienna in the early evening.

Airmail could be addressed to other major cities of the Empire. It would be forwarded from the nearest airmail terminus as special delivery. Such mail is quite scarce. I have seen a Vienna–Temesvár cover that was flown to Budapest; and a Budapest–Prague cover flown to Vienna is illustrated later.

Airmail Quantities. Since this airmail service operated for only three weeks, many collectors attempt to find a piece of mail from each flight. It is a worthy and extremely challenging goal. With this goal in mind, we should first address the quantities of mail carried each day. The table below provides this data. Note that the

Mail Carried on the 1918 Budapest–Vienna Airmail Service

DATE	Bp to Vienna		to Cracow		to Lemberg		to Bp
July	Norm	Regd	Norm	Regd	Norm	Regd	Norm
4	520	0	0	0	0	0	NF
5	651	431	0	0	0	0	263
6	109	169	0	1	0	5	174
7	68	69	0	0	0	0	170
8	428	206	0	2	0	2	56
9	461	249	0	1	1	4	227
10	436	191	306	4	0	5	170
11	181	125	70	0	44	0	177
12	118	124	0	0	26	3	165
13	36	97	0	0	0	0	122
14	95	34	1	0	0	0	117
15	120	100	1	0	73	1	47
16	36	59	4	1	0	1	109
17	306	95	0	10	13	2	93
18	37	60	0	0	0	1	100
19	43	217	0	0	0	3	124
20	88	65	0	1	0	2	89
21	11	0	0	0	0	0	70
22	12	66	1	0	0	6	NF
23	39	103	0	0	0	0	67
24	NF						71
	----	----	----	----	----	----	----
Total	3795	2452	383	20	157	35	2411

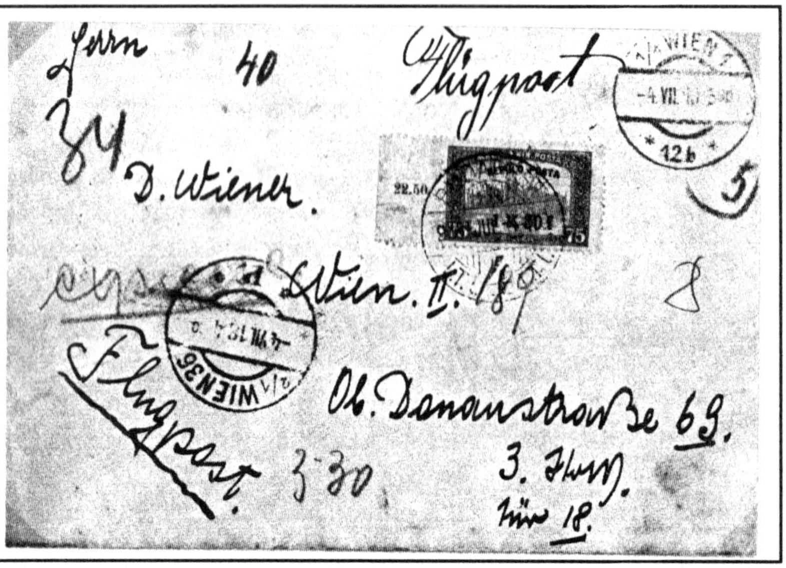

Figure 3–65 First Flight Cover Budapest–Vienna

numbers indicated for Cracow and Lemberg are only for the mail originating in Budapest. Mail originating in Vienna is much more common on those legs of the airline. **NF** indicates No Flight was made in that direction on the given day. Also, see *Budapest–Kiev Mail* for more data.

The table shows the number of pieces of normal and registered mail carried on each flight between Budapest and Vienna. Under 7,000 total pieces of Hungarian airmail were flown, with mail flown to Cracow and Lemberg being quite scarce. Probably less than half of all the mail survives.

The rightmost column in the table lists mail flown from Vienna to Budapest. Such mail is *much* scarcer than Vienna–Cracow or Vienna–Lemberg airmail. Austrian regulations did not permit registered mail to be carried on the flights originating in Vienna.

First Day Airmail. For the first flight, Hungarian airmail was canceled with a normal Budapest double–ring *cds* with the counter mark *LTl*. A second counter mark *LFe* is seen less frequently on first day mail. In the past some have questioned the legitimacy of this second marking. But, it is now generally recognized as being genuine.

Figure 3–65 shows a first flight cover from Budapest to Vienna. Additional franking of 1K80f in regular stamps is on the reverse. The Vienna airmail receiving mark is of July 4 at 1:40. [Note: on Austrian receiving marks, the hour in Roman numerals indicates AM; in Arabic numerals PM.]

Many specialists now believe one entry in the quantities table is incorrect, and that registered mail *was* carried on July 4. The total amount of mail doesn't change; registered mail was simply not counted separately the first day.

The lowest registry etiquette number found to date on a genuine flown cover is *4004/r*. Actually, expect registry etiquettes for this entire airmail service to be in the *4000/r*–series. [Note: etiquettes were not necessarily used in order, so ranges of etiquette numbers do *not* correspond to dates.] Most registered items flown were official mail and dispatches that never became available to philatelists. Therefore, registered mail from this service is scarcer than would be indicated by the table.

In 1973, Otto Schaffling and I wrote an article about these Budapest–Vienna first day covers. In it we gave guidelines on how to identify fakes. The bottom–line rules are the following:

1) The airmail stamps must be genuine (see *Forgeries and Fakes of the 1918 Issue*).

2) The cancellations and receiving marks must be as indicated above. At least one FDC, that otherwise appears genuine in every respect, is without a backstamp. It seems an occasional backstamp was missed.

3) The franking must be correct. For a normal weight letter to Vienna this is the 1K50f airmail stamp and 1K80f in regular postage.

4) You should find three manuscript markings on each cover. They are the word "expresse" usually written in blue crayon; a penciled

Figure 3-66 First Flight Vienna-Budapest

number that is a running count of the mail processed that day; and a penciled indication of the total postage required.

The word *expresse* must be there on first flight covers. Look at the example in Figure 3-65. The handwriting you see there is on many of the genuine covers canceled *LTl*. [Note: *expresse* -- i.e. special delivery -- service was required on all airmail; after the first couple of days, the mail was usually not explicitly marked with the word "expresse."] The penciled marks may have been erased by a well-meaning collector "cleaning up" his cover. Actually, the postage indication may even have been obliterated by the stamps themselves. Those are the guidelines; but actually, most fakes are poorly done and immediately obvious.

The other half of first-flight collecting on this route is July 5 mail Vienna to Budapest. These covers are *very* hard to find. Figure 3-66 shows an example. This mail was backstamped on receipt at *Budapest 4* at 9 AM.

Budapest-Vienna Airmail. On July 5, the *LTn* canceler was used on all mail. Beware of that date, especially if the cancellation is distinctly blue. Such cancellations were extensively used to create fake covers.

From the 6th to the 15th, the canceler used on all the airmail was *LLz*. Figure 3-67 shows an interesting cover with the *LLz* canceler. It was a heavy letter (you can still see the "stretch marks" in the envelope that were caused by the contents). It is correctly franked for triple weight (41-60 grams). It was aboard the ill-fated Budapest-Vienna flight of July 13, in which Lt. Varga was killed. Note also that it is addressed to Prague. It bears both a Vienna and a Prague backstamp.

Beginning on July 16, a special airmail postmark with *IND* (the abbreviation for *inditás*, meaning "dispatch") was used to cancel the airmail stamps. From the 16th, Budapest also used a special airmail receiving mark with the letters *ÉRK* (for *érkezett*, meaning "receipt").

A word is needed about the mail of July 16. On that first day of use for the new airmail

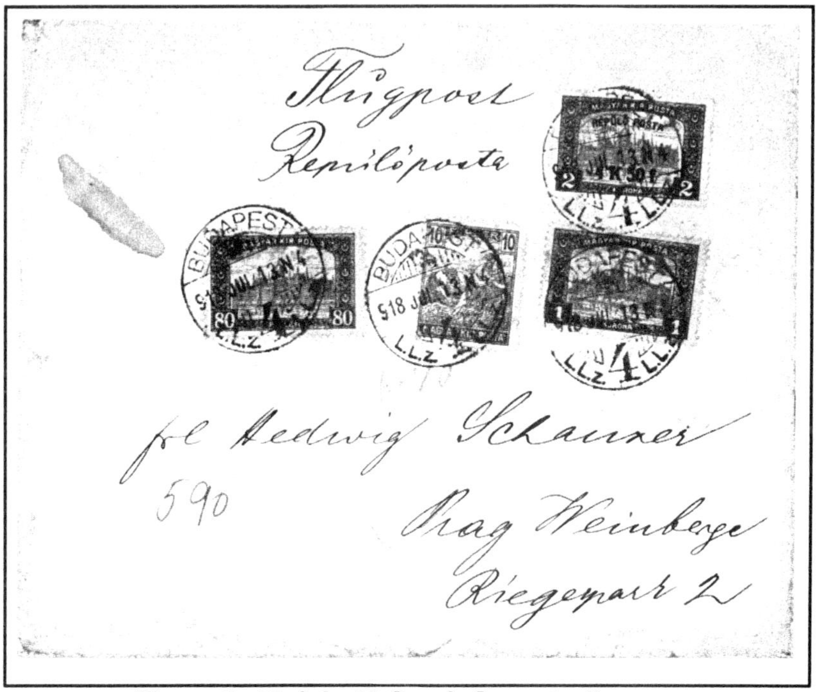

Figure 3-67 July 13 Crash Cover to Prague

Figure 3–68 July 23 Last Flight Budapest-Vienna

postmark, the rules were apparently unclear. You can find mail with all the stamps canceled by the *LLz* postmark that day, as well as mail with all the stamps canceled by the new *Repülö Posta / IND* marking. The correct procedure, which was used subsequently, was to cancel the airmail stamp with the *Repülö Posta* postmark and all the other stamps with the *LLz* postmark. This is how you will find all the airmail of July 17–23.

Figure 3–68 shows a registered cover of July 23. It was flown on the last westbound flight. It also illustrates the special *Repülö Posta / IND* postmark used since July 16th. Figure 3–69 shows a postcard flown from Vienna to Budapest on July 21. It illustrates the special *Repülö Posta / ERK* receiving mark, also used since the 16th.

Some collectors will have concern that the Vienna receiving marks on their Budapest–Vienna airmail are correct. The arrival postmarks were normally applied between 7 and 9 PM the day of the flight. I know of four exceptions. The July 5 airmail arrival was at 3:30 AM on the 6th. July 6 arrival was 9:15 PM on the 7th. This is after the mail of the 7th arrived. July 13 mail arrived at 11:15 AM on the 14th; and the July 17 arrival was at 6:30 AM on the 18th. I have never seen the reverse of a Budapest–Vienna cover of the 21st and so cannot confirm the arrival time.

Budapest–Cracow Airmail. Budapest–Cracow would have been the rarest of the 1918 airmail if it weren't for one Stefan Stiasny. He sent most of the mail carried from Budapest–Cracow to himself. Needless to say, he was philatelically inspired. The Stiasny mail is all dated July 10 or 11. He was a pilot on the airline, so quite possibly he flew some of the mail himself. His mail was correctly franked for a heavy letter. It is impossible to tell whether his letters actually weighed more than 20 grams. In any case, the heavy letter rate permitted him to show both of the 1918 airmail stamps on his covers to meet double the two-leg airmail surcharge. Figure 3–70 shows an example of a July 10 *Stiasny* cover to Cracow.

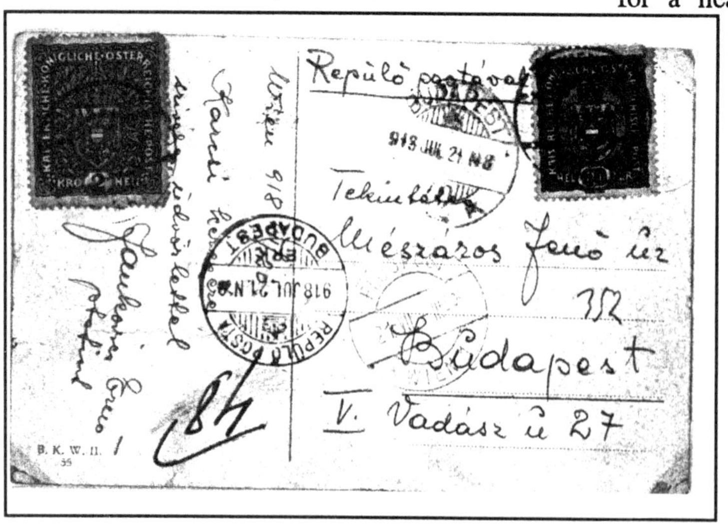

Figure 3–69 July 21 Vienna-Budapest Card

Remember that the mail arrived in Vienna from Budapest in the evening. The receiving mark was applied to the back of the envelope. Early in the morning, before take-off for Cracow, a dispatching postmark of Vienna was applied to

Figure 3-70 July 10 Budapest Cracow *Stiasny* Cover

the front. On arrival, a Cracow receiving mark was applied. So, correctly marked mail should have a Hungarian cancellation and Vienna dispatch mark on the front, and a Vienna and Cracow receiving mark on the reverse. It is not infrequent that one or the other of the Vienna markings is missing.

Figure 3-71 shows one of the very few examples of Budapest-Cracow airmail which is not addressed to Stefan Stiasny. This is a correctly franked postcard of July 16, which uses two copies of the 1K50f stamps for the two flight legs. It shows the *Repülö Posta / IND* postmark canceling all the stamps. Both the Vienna arrival and dispatch marks are on the front of the card. The Cracow arrival marking is on the back.

Most sources show no record of airmail to Budapest originating in Cracow or Lemberg. It is logical that some existed. Figure 3-72 shows an example. It is a cover flown from Cracow on July 8, arriving in Vienna at 8 PM. It also has a July 9 early AM dispatch postmark of Vienna for the flight to Budapest. Unfortunately, there is no Budapest receiving mark on it. It appears genuine, and I believe it is just another case of a missed receiving mark.

Budapest-Lemberg Airmail. Though Stiasny sent a few pieces of airmail to Lemberg, he didn't do for that city what he did for Cracow. Therefore, Budapest-Lemberg airmail is the scarcest of the normal airmail destinations.

The cover shown in Figure 3-73 was flown from Budapest to Lemberg. It flew Budapest-Vienna on July 10. It is missing the Vienna dispatch mark, but on the reverse it has three arrival marks: a Vienna airmail arrival of the

Figure 3-71 July 16 Budapest-Cracow Postcard

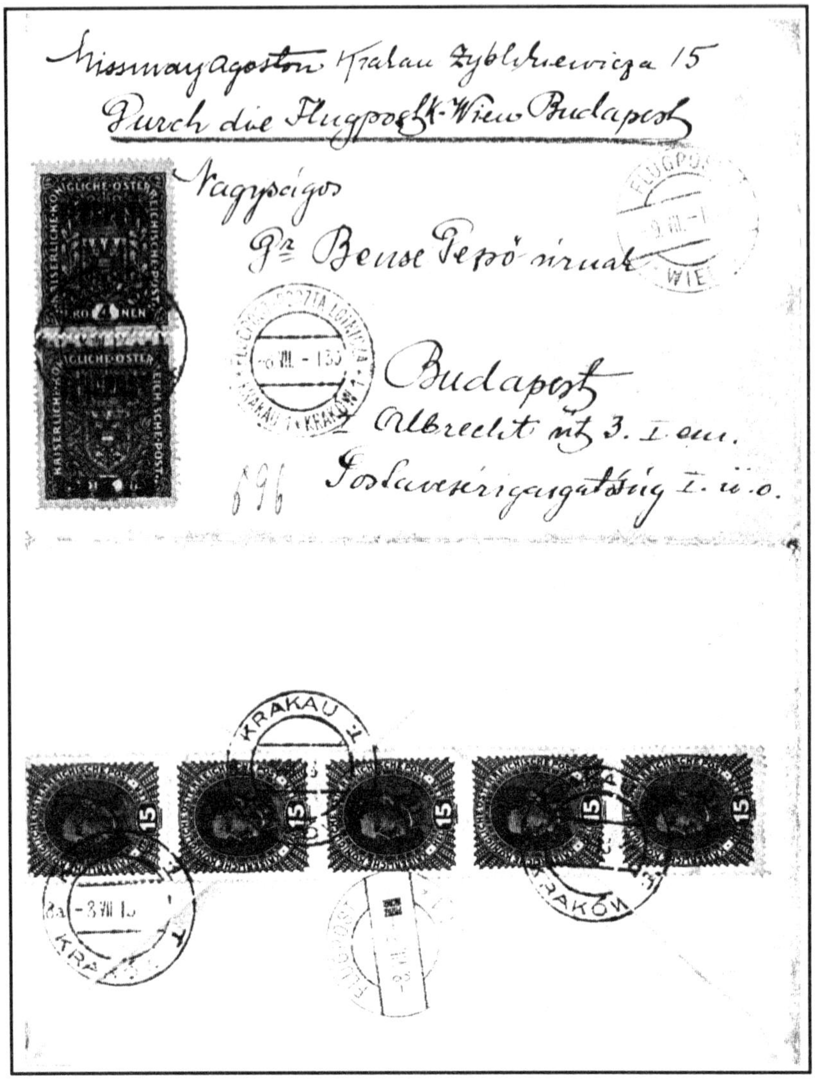

Figure 3-72 Cracow- Budapest Airmail Cover

10th, a Lemberg airmail arrival of the 11th, and an *Ettapenpostamt 185* receiving mark of the 13th. This "base postoffice" presumably was in the field at that time. It was a major headquarters field postoffice that, according to Tranmer, was located in Lemberg in 1916. I do not know where it was located in 1918, when the war on the eastern front was over.

Budapest–Kiev Airmail. I do not consider Kiev a normal airmail destination for the airmail service under discussion, since normal civilian mail was not permitted beyond Lemberg. The Lemberg–Kiev route was limited to official and military mail. There are experts who insist that a few (less than five) genuinely flown pieces do exist.

I have personally examined only one of these. It is from a government office in Budapest to an official delegation in Kiev. It clearly was flown from Budapest to Vienna on the July 6 flight, and it certainly arrived in Kiev where it was censored. But, it is registered. Registered mail was not permitted on the airline Vienna–Cracow–Lemberg–Kiev. There is also no Vienna airmail dispatch mark, and no airmail arrival mark other than the Vienna mark applied on its arrival from Budapest. It is a wonderful cover, and *is* flown airmail. It went from Budapest to Kiev, but I doubt that it flew the *entire* distance.

Budapest–Kiev airmail is so rare that each item surely has a story all its own. It would be interesting to try to gather pictures of each of the known covers, and write those stories. I'll leave that task for another author.

Air Station Budapest. When the 1918 airmail service terminated, the Army aviation presence in Budapest did not disappear. Things simply went back to the *status quo* of three weeks earlier. Civilian airmail was not handled. I wouldn't be surprised if Capt. Raft–Marwill had the opportunity to say "I told you so."

Figure 3–74 shows a September 23, 1918 official cover with the very unusual two–line boxed unit marking of the "Military Airline Station Budapest" at Mátyásföld. Whether or not it was flown is indeterminate. By my own rules, that means it probably wasn't. It is still *very* rare. Until the war was over, *Air Station Budapest* continued to operate.

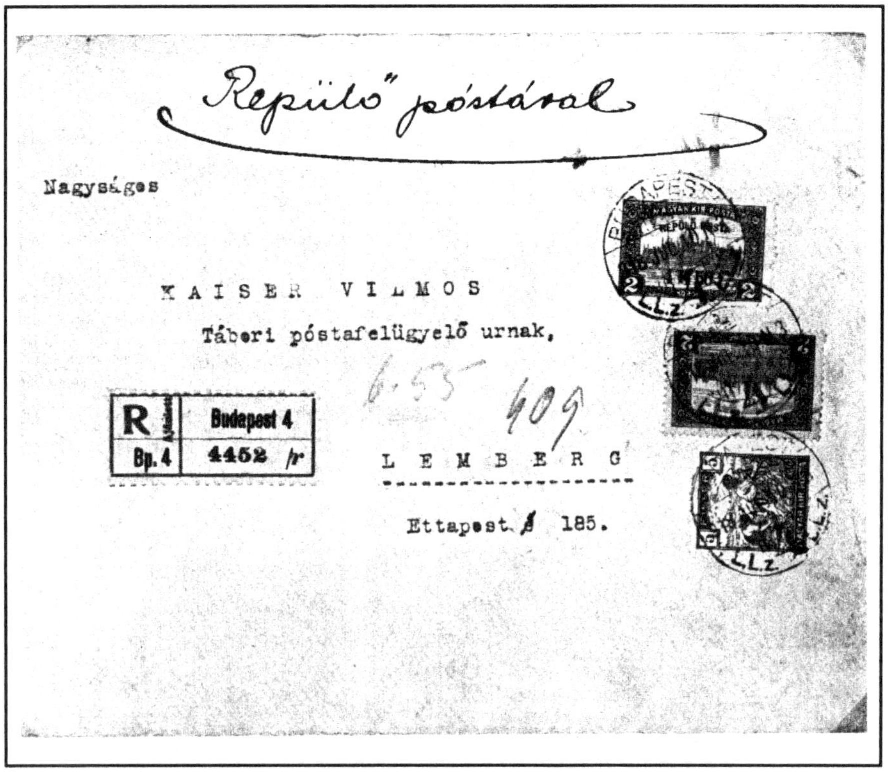

Figure 3–73 July 10 Budapest–Lemberg Airmail

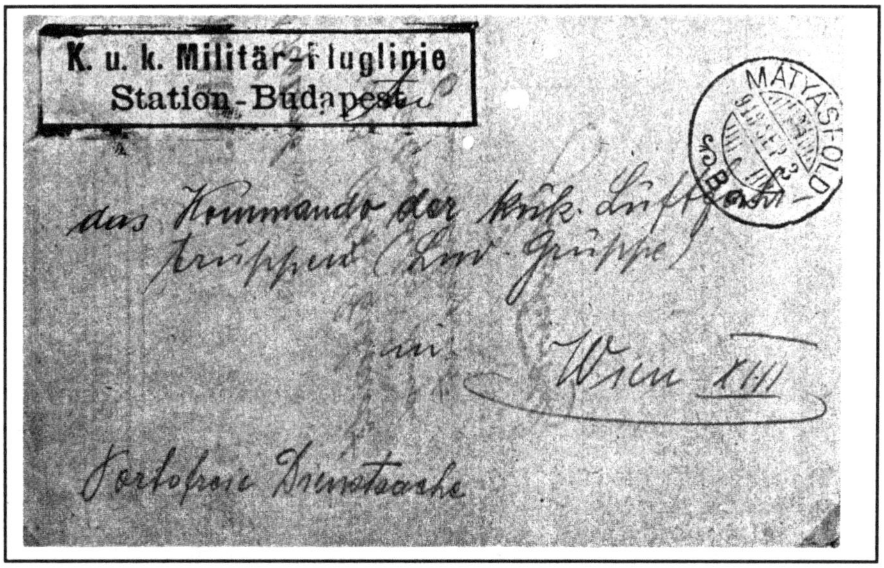

Figure 3–74 Air Station Budapest Unit Marking

Forgeries and Fakes of the 1918 Issue

Where there is profit to be made, somebody will find a way. Because the airmail stamps of 1918 were overprinted on what are very inexpensive Hungarian stamps, there was profit to be made by forging the overprints. Also, since flown mail is reasonably scarce, there is profit to be made in faking covers, possibly even using forged stamps. And finally, there is the issue of the variety of the 1K50f airmail stamp with the overprint inverted. We will discuss those matters here in that order.

Figure 3–75 Forged Overprints

Forged Overprints. The overprinting of the airmail stamps of 1918 was a rush job, in war–time. Therefore, you would expect, and you will find, considerable variation in the quality of the impressions.

Even so, it is relatively easy to spot the forg–eries. There are four keys to look for, and they are related to four characters in the overprint of both stamps ... the *K*, *5*, *0*, and the *f*. The characteristics are shown –– *very exaggerated* –– in the drawing of Figure 3–75. The characteristics in the figure marked "G" represent the genuine overprints, those marked "F" represent the forged overprints.

In the genuine overprint, the *K* has serifs on its "legs" and "arms." On the forgeries, the serifs are either weak or non–existent.

The curved top line and the short vertical line of the *5* are distinct in the genuine overprint. In the forgeries they merge together to form a triangle.

The oval forming the *0* is thinner at the top and the bottom of the genuine overprint. The forgeries have a more constant thickness all around the oval.

The top of the *f* forms a distinct hook in the genuine overprint; the forgery has just a blob of color where the top of the *f* should be.

The presence of *any* of these indicators often means you have a stamp with a forged overprint. More than one indicator makes it a virtual certainty the overprint was forged.

Faked Covers. If the stamp was forged, obviously the cover was faked. But, there *are* faked covers with real stamps. Figure 3–77 actually has an example of both on a single cover.

Look for proper franking and proper postal markings. These stamps could be used nowhere but Budapest, and the mail had to be addressed to Vienna or another major *Austrian* city –– not to Budapest or any other Hungarian city. See Figure 3–77 again for an example of this problem.

Finally, the dates of the cancellation must be July 4 to July 23, 1918. Any postmark outside that range is an obvious fake.

As noted above, there are many fake covers with a blue cancellation of July 5. Also, beware of intense, shiny black cancellations. It was wartime, and the ink wasn't that good.

The Inverted Overprint. In 1919, one sheet of the 1K50f airmail stamp was "discovered" with the overprint inverted. For many years this variety, which could obviously never be found on a genuine cover, was listed by the Scott catalog as C1a ... usually at a good price for the time. Profit was to be made, so these 100 stamps quickly turned to thousands as forgers copied the inverted overprint.

There is a second question ... what is the provenance of that original sheet? I believe it was a "fake." Note that I'm not using the word forgery –– you can't forge something that doesn't exist! As a result of my efforts, and at the official request of the Society for Hungarian Philately, Scott "delisted" C1a in 1987. Whether or not you agree with me, or consider the original sheet of the inverted overprint airmail to be genuine, this delisting served a useful purpose. It drove off the market the obvious forgeries that were intended to copy the original 1919 sheet of inverted overprints.

Before I attempt to deal with my conclusions on the original sheet, I'd like to address the obvious forgeries. Many forgeries have the same characteristics as the forged upright overprints. That should be obvious, since it is no more difficult to apply a fake overprint upside–down. But, there is also a more insidious forgery that was prevalent on the American market in the 1970's. This is what I call the "tall 0" variety.

The "tall 0" variety was created to circumvent a problem identified with the original sheet of inverts in 1925. It has the peculiar characteristic of a specific cliché within the sheet, that was supposed to "guarantee" it as genuine. Of course, since there was only one sheet of the inverts known, there could be only one stamp from this position. It didn't take long to realize there were altogether too many examples of this one plate position. Yet they continued to be sold to the unwary. The characteristic in question is that the umlaut over the *u* in *Repülö* is misformed, making the letter look like an *0* instead of a *ü*.

Figure 3-76 C1a Forgery -- the "Tall 0" Variety

[Note: the partial APS expertization certificate shown in Figure 3-76 correctly declared the stamp pictured to be a forged overprint.]

Having dealt with the obvious forgeries, let's discuss the facts about the original sheet of inverts found in 1919. I believe, based on the opinions of four of the greatest experts of Hungarian philately, that the original 1919 sheet was a fake. That means a genuine airmail inverted overprint error never existed. Many very reputable experts disagree. I respect their right to their opinions. Since my interest and area of expertise is aerophilatelic mail, not stamps, for my collection and me this is an academic issue. Here are the facts.

In the March 1925 issue of **Philatelia** (Vol. 1 No. 4), Franz Doszkár, a noted Budapest stamp dealer and philatelic author, published an article titled "Forgeries of the Hungarian Air-Post Stamps, First Issue, with Inverted Surcharge." This article discusses the origin of the original 1919 sheet of the stamps we knew as Scott C1a. It documents the differences of opinion concerning their legitimacy that were current in the mid-1920's. It also details his personal research into five stamps from the 1919 sheet (a strip of three and two singles) to which he had access. Doszkár's conclusion was "*Kurz und gut, die Flugpostmarken mit verkehrtem Aufdruck sind wertlose Fälsch-ungen.*" The translation from the German is: "Short and sweet, airmail stamps with inverted overprints are worthless fakes."

He notes that the first mention of the inverts was in the February 5, 1919 issue of **Donaupost**. The announcement claimed the sheet of inverted overprints of the 1K50f airmail stamp of 1918 was the "greatest rarity of wartime Hungarian stamps." It noted that the sheet had been divided into blocks of 50 and 40, and a strip of 10. The announcement offered the block of 40 for sale at 15,000 Korona.

Doszkár never mentions the name of the owner in the article cited. It was widely known that the owner was the rather disreputable dealer Taussig, and Doszkár mentions him as the owner of several of these stamps in later articles. I have in my possession unverifiable personal correspondence that claims Taussig sold the block of 50 in 1920 to the Vienna dealers, the Marbach brothers.

Doszkár devised an experiment using the five stamps mentioned, since they had identifiable positions in the sheet. Position is determined in sheet margin copies by selvage markings. He carefully compared the five stamps to the corresponding positions in the reference sheet, with upright overprint, in the archives of the

Hungarian Postal Museum. His most telling argument came from comparing positions 11, 12, and 13 in the reference sheet to the strip of three that were identifiable as positions 90, 89, and 88 in the inverted overprint sheet. [Note: since the sheet was inverted prior to printing the surcharge, the same overprint clichés were used for printing the two strips.]

He noted that the total end–to–end width of the three overprints on the reference sheet was 81.5 mm, while on the inverted overprints it was 79 mm. Therefore, he proved the inverted overprint was printed with a different plate than the one used by the Hungarian Government Printing Works in 1918. If you'd like to see for yourself how the settings vary, measure the end–to–end width of the overprints in Figure 3–62. The stamps are shown actual size. You'll see that the width of the leftmost three overprints is greater than for overprints taken three–at–a–time further to the right.

An interesting secondary effect of Doszkár's analysis came from the fact that he noted the reference sheet had a distorted umlaut in position 31 that was not found on one of his single stamps -- i.e. position 70 inverted. I'm sure his statement about this peculiarity of the reference sheet led to the unscrupulous creation of the "Tall 0" variety.

In a follow–up note later in 1925, Doszkár writes that "Rudolf Friedl of Vienna notified me that Mr. Taussig exhibited fake copies [of C1a], and that he [Friedl] had declared them counterfeit." This establishes that two of the greatest philatelic experts of that time agreed that Taussig's 1919 sheet was a fake.

There is another side to this argument. After denials for a decade, the Hungarian Printing Works granted full recognition to the sheet of inverts in the late 1920's. Since then, the official establishment in Hungary has supported the position that the original 1919 sheet of inverts are genuine. There is a statement to that effect in the "bible" of Hungarian philately ... **A Magyar Bélyegek Monográfiája**.

Unfortunately for that official position, *Philatelia Hungarica* (then the government–owned philatelic distributor) published a photo in its 1971 English–language catalog of the inverted airmail overprint. It was clearly the "Tall 0" variety that everyone knows is a fake! By the way, they priced it at US$500 *on cover*! It was just that kind of unconscionable greed at which I focused my 1985 research on this subject.

In 1985, I also looked for current expert opinion on the subject. Mr. Herbert Bloch, the outstanding philatelic expert of our time, commented on the subject. He indicated that he considered the 1925 Doszkár article to be the "definitive statement on the 1918 airmail inverted overprint." He also said he did not "recall ever signing an example of this stamp as genuine."

Mr. Otto G. Schaffling, a noted expert on Hungarian philately who recently passed away, reviewed this controversy in 1985 at my request. He stated the following: "During this period [1919–20], several print shops in Budapest forged overprints of the Occupation issues. Many of these forged overprints were sold by Taussig. He was not a reputable dealer. Doszkár was one of the most reputable dealers in Hungary at that time, and I believe his *Philatelia* article proves the 1919 sheet of C1a to be fake. That Taussig 'found' the sheet with inverted overprint makes it immediately suspicious. Therefore, I conclude that the invert sheet was initially faked, probably at Taussig's order."

I stand by my 1985 article. Scott and many U.S. collectors of Hungarian stamps stand by it as well. I still believe that the original sheet of 100 airmail stamps with inverted overprint was faked.

Figure 3–77 An Outrageous Fake!

Figure 3–77 shows a truly "fun" cover. It is perhaps the prettiest *and* most outrageous fake I have ever encountered. It is a July 20, 1918 "airmail" cover that is franked by a fake copy of the inverted overprint airmail (not the "Tall 0" variety), by a genuine copy of the 4K50f airmail of 1918, and by 41 fillér of regular stamps. It is addressed to Budapest, and has no backstamps. It uses the terminology *Légiposta* for airmail, which wasn't used until 1920. An honest dealer *gave* this cover to me as a "thank you" for my writings on the 1918 airmail issue of Hungary. It also issues us all a warning. Someone obviously had access later to either the original *Repülö Posta / IND* canceler or a very good facsimile.

Budapest: A 1920's Airmail Hub

This section of the book focuses on the decade of the 1920's. It deals with both domestic and international airmail. An abortive attempt to establish domestic airmail routes in 1920–21 was ended by a strict interpretation of the Treaty of Trianon. Later in the decade, domestic airmail was limited to a series of philatelicly oriented events.

By 1922, foreign and Hungarian business interests recognized that Budapest was a vital link in the expanding European airline network. Geography, and the range of 1920's aircraft, demanded that Budapest be an airline hub, and airmail was key to funding the aviation industry of the 20's. The principal elements of studying airmail routes, rates, and postal markings are very intertwined in the 1920's saga of Budapest airmail. I've tried to sort these issues out into three chapters, but didn't succeed in fully separating them. I believe that from 1922–26 the Hungarian government stood almost alone in *discouraging* the use of airmail. I will explore this enigma in the chapter on "rates." Throughout, I hope to show the marvelous collecting opportunities offered by the international airmail of the 1922–27 period.

Domestic Airmail Service of 1920–21

Throughout Hungary's aerophilatelic history, *speeding the mails* was never the reason for maintaining a domestic airmail service. Hungary is a small country. Given the distances involved, land transport generally could -- and still does -- deliver the mail more quickly, regularly, and cost effectively than airplanes. Over the years, the real reasons for domestic airmail included a variety of political motivations, perceived status symbols, the promotion of tourism, and covert military activity. Most often, domestic airmail was simply philatelic. This was largely the case even in the 1920's.

In January of 1920, intending to establish an air force, Admiral Horthy's government purchased in Vienna eight Phoenix Type C–I biplanes with 230 hp Hiero engines. On February 11, 1920 the Commerce Ministry authorized the formation of the *Magyar Aeroforgalmi RT (MAEFORT* or in English the "Hungarian Air Traffic Company") to provide domestic aviation services. Count László Károlyi led the company, the capitalization of which was 50 million Korona. The government planes were transferred to MAEFORT, which had 12 more Phoenix biplanes built at the Hungarian Lloyd airplane factory, and had smaller planes built at other Hungarian factories.

The company established facilities in Budapest (operating out of the Mátyásföld airport), in Szombathely (which was the company's maintenance and pilot training center), Miskolc, Szeged, and Székes-fehérvár. During 1920–1, MAEFORT operated 28 airplanes using former military pilots, and received significant government subsidies. It is no wonder that Czecho-slovakia soon accused the company of being a *front* for military aviation activities that violated the Treaty of Trianon.

Figure 4–1 The Airmail Issue of 1920

Figure 4–3 MAEFORT Planes on *Rákos Flying Day*

The first "postal service" provided by MAEFORT was the delivery of 1600 newspapers from Budapest to Szeged on May 12, 1920. The first real *airmail* service began with an air show at Rákos, Budapest's older airport. Airmail stamps listed by Scott as C3–5 were issued for this new domestic airmail service. These are the only Hungarian stamps specifically issued for domestic airmail. Figure 4–1 shows the high value of the *Airmail Issue of 1920*.

The 10 K value of the then current Parliament series was overprinted *Légi Posta* and the denomination. The three values, 3K, 8K, and 12K, were intended respectively for newspapers, postcards, and letters. The colors of the overprint were green, red, and blue. No imperfs exist. Over 300,000 copies of each were printed; so these stamps remain inexpensive in unused condition.

A double–overprint of the 8K value exists. Figure 4–2 shows a genuine example of the double–overprint variety. Note that the second overprint is rather light, some would say "just a kiss," and it is upward and to the right of the normal overprint. This variety was unknown until after WW II. It may be from printer's waste stolen at the end of the war. Formerly this double–overprint was listed by Scott as C4a. It continues to be listed in Surányi's *Handbook*.

The *Rákosi Repülö Nap* ("Flying Day at Rákos") took place on Sunday, November 7, 1920 and featured a variety of events and competitions. A highlight of the day was three Phoenix biplanes taking off for Szombathely with what was termed the first Hungarian domestic airmail. It was intended as a precursor of a regular Budapest–Szombathely airmail service. The photo in Figure 4–3 shows MAEFORT planes on the flight–line at Rákos. As will be noted later, planes also went to other provincial cities from the event, without airmail.

The mail posted for the first flights bears a fancy oval marking reading "*Airmail Inaugural / 1920 Nov 7 / Flying Day at Rákos.*" Some of the mail also bears a variety of vignettes. Figure 4–4 shows an example flown to Szombathely. It illustrates the *Rákosi Repülö Nap* marking on a cover with a typical vignette.

Figure 4–2 Double Overprint on 8K Value

Mail was also dropped by parachute to Györ and Pápa from the first flights of November 7. The receiving marks are both at 3 PM that day. Due to poor weather, after the mail drops all three planes made forced landings and did not arrive in Szombathely until the next day. The mail carried to

Figure 4–4 Budapest–Szombathely First Flight Airmail

Szombathely by each of the airplanes is distinguishable by the receiving mark. The three receiving marks known are November 8 at 9 AM and 1 PM, and November 9 at 10 AM. The reason for some mail being delayed an additional day is unknown. The receiving mark of the 9th is comparatively scarce.

Figure 4–5 shows a first flight cover that was dropped to Györ. The Györ mail is often, but not always, handstamped with a small circular military censorship mark as seen in this figure. Figure 4–6 shows a first flight cover that was dropped by parachute to Pápa.

The airmail service to Szombathely was to operate three times each week -- on Monday, Wednesday, and Friday. [Note: though flights were scheduled for those days, mail could be postmarked on other days of the week.] With large

Figure 4–5 First Flight Drop Mail to Györ with Censor Mark

Figure 4-6 First Flight Drop Mail to Pápa

quantities of mail having been flown on the 7th, there was little mail for the first *regular* flight of the 8th. In fact, though, many postcards addressed to Györ were flown on the 8th. They bear the oval cancellation of the *7th*, and one of several MAEFORT company handstamps. These were flown by the airline for philatelic resale. Figure 4-7 shows an example of such a card.

Also canceled and flown on the 8th was a small quantity of other mail -- for unknown reasons most of it is addressed to Györ. Figure 4-8 shows an example.

An interesting research finding is that the Györ mail of the 8th was dropped on the *return flight* from Szombathely to Budapest. This mail has a 10 PM Györ receiving mark. A very unusual November 7 cover addressed to Györ proves that the Györ mail drop took place on the return flight of the 8th. The letter in question, shown in Figure 4-9, went mistakenly to Szombathely. It shows November 8 receiving marks of both Szombathely at 9 AM and Györ at 10 PM. Note that the Györ receiving mark is identical in Figures 4-7, 4-8, and 4-9. The only known instance where mail was dropped from a

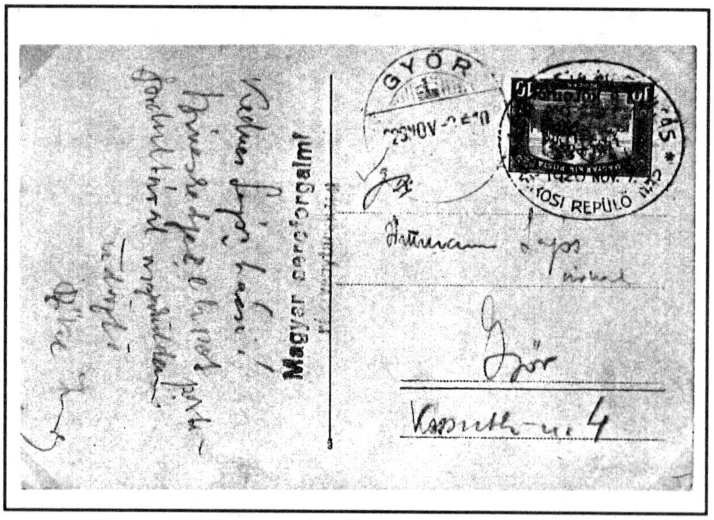

Figure 4-7 MAEFORT Philatelic Card Flown on the 8th

Figure 4–8 November 8 Postcard to Györ

Szombathely to Budapest return flight was on November 8.

The first flight in the Szombathely–Budapest direction was on November 8. Figure 4–10 shows an example of first flight mail in the eastbound direction. Airmail stamps canceled–to–order in Szombathely on November 7 exist, but the earliest postmark on flown mail is November 8.

Two other very scarce airmail varieties of November 8 exist. This is mail apparently flown to Budapest from Szeged and Debreczen on that date. Hozér and other experts believe MAEFORT airplanes went from the Rákos Flying Day event to these provincial cities without airmail. On the following day they returned carrying mail unofficially.

Since regulations required that airmail stamps only be canceled by an airmail marking, this unofficial Szeged and Debreczen airmail was handled in an unusual way. The mail was postmarked at the origin, but the stamp(s) were not canceled there, since airmail cancellers did not exist at that time in either city. The airmail stamps were canceled on arrival in Budapest. Figure 4–11 shows an example of a postcard from the November 8 Szeged–Budapest flight. Don't confuse this November 8 mail with the regular, official airmail service between

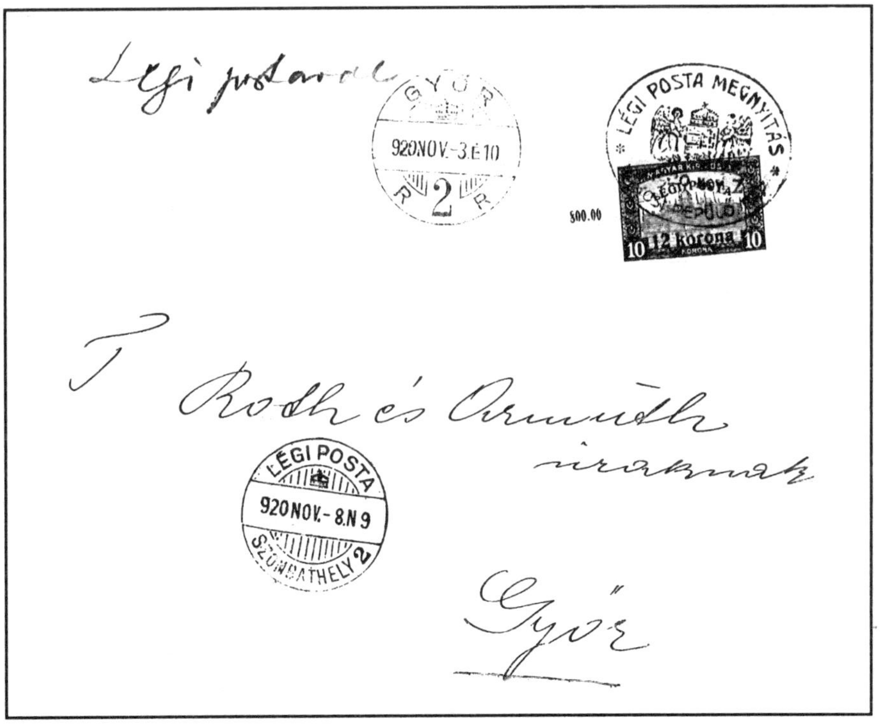

Figure 4–9 Airmail with Györ *and* Szombathely Receiving Marks

Figure 4-10 First Flight from Szombathely-Budapest

Figure 4-11 Nov 8 Unofficial Szeged-Budapest Airmail

Szeged and Budapest inaugurated in December 1920.

Following the first flights, and through the winter of 1920–21, MAEFORT flew the route between Budapest and Szombathely intermittently, depending on the weather. Airmail flown in both directions, on a variety of dates, is not unusual. [Much of the mail known bears November or March dates, indicating that during the worst winter weather many flights were canceled. Alföldi and Talas attempt to provide a composite list of actual flight dates.]

Mail flown to Szombathely and then forwarded to other towns shows a Szombathely receiving mark. Mail with other receiving marks, but without a Szombathely mark, confirms that mail drops were made to those towns. After November 7–8, most of the drop mail was to Györ and Sopron. Other cities confirmed as receiving mail drops on later dates include Pápa and Köszeg.

Figure 4–12 shows an example of the rather common drop mail to Sopron. Figure 4–13 shows an example of the very unusual drop mail to Köszeg.

Other probable drop mail sites include Veszprém, Zalaegerszeg, and Herczeghalom. Mail to these towns usually bears no receiving mark, indicating that it wasn't flown to Szombathely. But, receipt by air at these destinations is not confirmed.

Postal regulations required that an airmail canceller must

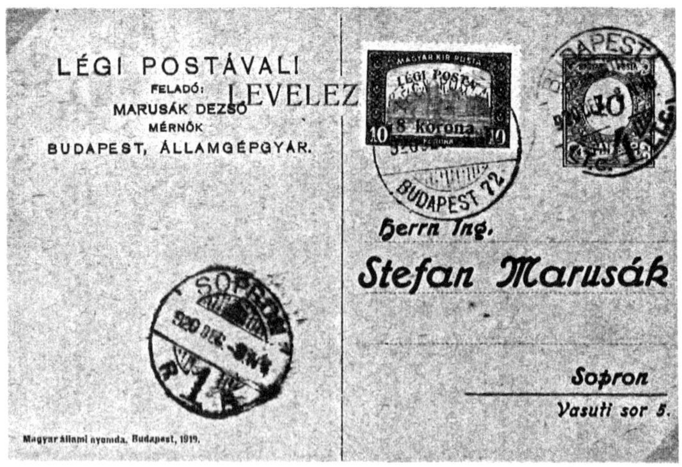

Figure 4-12 Budapest-Sopron Drop Mail

Figure 4-13 Drop Mail to Kőszeg

obliterate airmail stamps. This rule created some unusual pieces, such as the November 8 mail mentioned above. Also, airmail originating at Sopron is not uncommon. Sopron *airmail* was postmarked at the origin and forwarded to Szombathely by auto. There, the airmail stamps received the Szombathely airmail marking. Such mail exists from several dates. Figure 4-14 is an example of a Sopron-Budapest "airmail" postcard.

On December 10, MAEFORT initiated a weekly airmail route between Budapest and Szeged. Airmail from Budapest to Szeged shows the *Budapest 72* airmail cancellation and a Szeged receiving mark. Most airmail originating in Szeged shows the Szeged airmail canceller and a Budapest receiving mark. Both receiving marks are normal circular date stamps. The first flight in the Szeged to Budapest direction took place on December 11. The Szeged post office accepted mail for this flight on the 10th and 11th; but, apparently the Szeged airmail canceller was not available until the 11th. Therefore, airmail of the 10th was postmarked with the normal Szeged *cds* and the stamps were canceled on receipt in Budapest. The mail of the 11th shows the new Szeged airmail cancellation. Few people used the Budapest-Szeged service, and so this airmail is very scarce. Figure 4-15 shows a March 1921 example of flown Szeged-Budapest mail.

Mail drops to intermediate cities on the Budapest-Szeged route may have occurred. Latham reported drop mail to Nagykörös and Kiskunfélegyháza from Szeged-bound flights. I have never seen or otherwise heard of such drop mail. If it exists, it is exceedingly rare.

Figure 4-14 Sopron-Budapest "Airmail"

Figure 4-16 shows a most unusual cover. It is airmail originating in Szeged that is addressed to, and backstamped at Székesfehérvár. This is possibly an example of the elusive drop mail from the Szeged-Budapest route. But, in my opinion, since MAEFORT had a landing field and other facilities in Székesfehérvár, it is more likely

Figure 4-15 Szeged-Budapest Airmail

the aircraft landed there on at least this one occasion with some mail. In any case, it is a most unusual cover.

Two surprising finds of Szeged airmail occurred recently. These are pieces originating in Szeged and addressed to Szombathely. They were apparently flown to Budapest and then on to Szombathely. They bear only a Szombathely receiving mark. There are no Budapest markings. These are the only reported examples of mail flown on both of the domestic airmail routes of 1920-21. The example shown in Figure 4-17 is a newspaper wrapper postmarked on January 16, 1921 and received in Szombathely on the 19th.

In early 1921, the Czechs filed a complaint with the Allied Control Commission. They claimed that the MAEFORT airmail operation violated the Treaty of Trianon. On April 9, the Commission ordered it shutdown. According to Steiner, who did unpublished research on the subject, the last mail flight took place on Wednesday, April 13 from Szombathely to Budapest. Figure 4-18 shows the only reported example of airmail of this date.

Further Allied actions against MAEFORT culminated in the destruction of all their aircraft in Budapest on December 6, 1921, at which time they obviously ceased all operations. This was the end of Hungary's first airline in the modern sense, and its first domestic airmail service.

Yet, this was not the end of the airmail issue of 1920. Examples of these stamps are found legitimately used on international airmail in 1922. Though they continued to be valid, the rapid increases in postal rates due to inflation resulted in them not being practical for on-going use -- their denominations were just too small.

Figure 4-16 Unusual Szeged-Székesfehérvár Airmail

Figure 4–17 Newspaper Wrapper Flown Szeged–Szombathely

Figure 4–18 Last Flight of the 1920–21 Airline

Then there is the one last, albeit somewhat illegal, use of the 8 K value shown in Figure 4–19. It is the kind of cover that makes our hobby lots of fun. The stamp is on an envelope canceled by a *Paquebot* cancellation in Aden (now in Yemen) in October 1935.

Mail posted aboard ship, bearing stamps of any UPU nation, must be accepted at any port–of–call. The details of that rule are that the stamps must pay the first class postage rate in the country indicated by the postage, and the stamps must still be valid. "Ah, there's the rub."

As with all the Korona–denominated stamps, the airmail issue of 1920 was invalidated on March 31, 1927. By that time, no less eight years later, their postal value had long since eroded from inflation. I suspect that neither the Anglo–Iranian Oil Company tanker captain that handed the letters over the counter, nor the Arab postal clerk were concerned with that kind of detail.

I found this "airmail paquebot" cover among a large group of similar covers all addressed to the same sea captain. Each was franked with stamps from a different country around the world. They numbered over fifty in all. All were canceled either at Aden or Penang, Malaya. I suspect the ship's captain was a philatelist! The joy of the search is philately's biggest benefit.

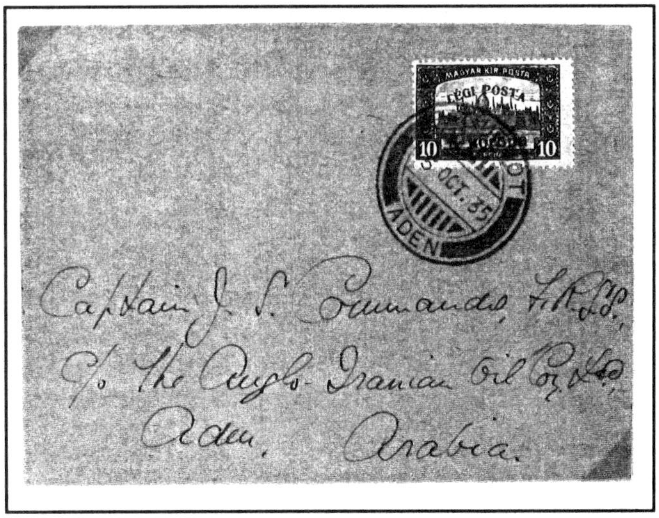

Figure 4–19 "Late Usage" of 1920 Airmail

Domestic Airmail Events of 1924–29

No domestic airmail was flown in Hungary during 1922–3. From September 20–24, 1924, Budapest celebrated the *Elsö Magyar Filatelista Nap* -- "First Hungarian Day of Philately." This series of philatelic events continues to the present, with only a short wartime interruption. For long, they have been known as *Stamp Days*. The highlights of this **1st Stamp Day** celebration were an excellent philatelic exhibition and round–trip airmail flights between Budapest and Hungary's ancient capital of Esztergom. Seaplanes, operated by *Aeroexpress*, *Junkers*' Hungarian affiliate, made these flights from the Danube.

A postcard with a blue drawing of Budapest, Esztergom and an airplane was prepared for the airmail flight of September 23. Figure 4–20 shows mail flown in both directions. Budapest mail was canceled with a *Magyar Filatelista Nap* roller cancel; Esztergom mail with a normal circular date stamp. Each also bears a directional boxed biplane cachet, franking of 10,000K, and a *cds* receiving mark. Most also have a 52 mm straight–line *Légipostaval Érkezett* (in English, "Received by Airmail") handstamp.

Only official postcards prepared by the exhibition committee were flown. The quantity flown is unknown; but, it was sufficiently large to keep the price of these cards low to the present day.

A second round–trip flight was made on September 25, the day after the celebration closed. Markings and franking are the same except for the dates. Presumably, the cards flown on the 23rd sold out, and a demand remained. On the 25th, in addition to the blue card, two other *Stamp Day* postcards were used–up for airmail. These are a green card picturing *Magyar Ritkaságok* ("Hungary's Rarities) and a red card picturing the 2,000K Parliament series stamp. The red and green cards used on airmail are much scarcer than the blue card.

Many of the postcards of the 25th remain unaddressed. It is likely that much of the mail on the first flight, and all on the second flight, was flown unaddressed for philatelic resale. Probably, the addresses seen on much of this mail today were added later.

Figure 4–20 ***1st Stamp Day*** **Budapest–Esztergom Airmail**

The *Elsö Bélyeghét Szegeden* -- "**First Stamp Week in Szeged**" -- took place in April 1925. The dates are indeterminate. Both the literature and the material give evidence of two different dates: April 7–14 and April 11–17. The round-trip flight between Budapest and Szeged was undoubtedly made on April 13.

Whether these flights are viewed as a special one-time airmail route, the reopening of the 1920-1 Budapest-Szeged route, or one of the commercial flights that regularly carried Budapest newspapers to provincial cities, is a matter of personal interpretation. While it is difficult to classify the flights, they certainly were *not* Hungary's "*First Inland Air-Mail,*" as is proclaimed by a small pink multi-lingual dealer's label sometimes found attached to this mail.

To help defray the costs of the exhibition, vignettes designed by Prof. Haranghy and inscribed *Aero Szeged-Budapest* were prepared for the airmail. The denominations are 1,000K for postcards and 2,000K for letters. The design is the same on both, but the colors differ. The postal authorities objected to the inscriptions of value. As a result, a second printing of the vignettes was produced in which a design resembling three zeroes replaces the indication of value.

The reprinting did not change the selling price of the vignettes. Though they were not required, a substantial percentage of the airmail has the vignettes affixed. Only the vignettes *without* value inscriptions were to be used. Figure 4-21 shows the original printing of the 1,000 K value. Figure 4-22 shows it as reprinted with the indication of value deleted.

Figure 4-22 Vignette as Used on Airmail

Along with the value inscription in the upper right, the two printings can be distinguished by color. As noted later, this can be very important in identifying genuine examples of an airmail rarity. The center of the 1,000K is a distinctly dull olive green, but the color is a bright green on the reprint. On the 2,000K, the frame is a bright brick red, while the frame of the second printing is a duller salmon red.

Postal order 8607/X authorized this airmail. The rates were 7,000K for postcards and 12,000K for letters. A boxed biplane directional marking was applied in violet in the Budapest-Szeged direction, and in black in the Szeged-Budapest direction. The plane left Budapest for Szeged at 11:30 AM carrying 4,389 cards and 3,486 letters. Upon receipt, a rectangular "Stamp Exhibition" marking, a *cds*, and the standard straight-line "Received by Airmail" handstamp were applied.

The return flight left Szeged at 3:30 PM carrying 5,400 cards and 2,414 letters. On receipt, the Budapest post office backstamped this mail with a normal *Budapest 72* airmail *cds* and one of three "Received by Airmail" marks. These were used because the normal handstamp was "mislaid." These markings all read *LÉGIPOSTAVAL ÉRKEZETT* and

Figure 4-21 Original Printing of 1925 Air Vignettes

Figure 4-23 Budapest-Szeged Airmail of 1925

were: (1) single-line, 86 mm long, in a rectangle; (2) two-line, 76 mm long; and (3) single-line, 120 mm long.

Most of airmail carried was cards and envelopes prepared by the exhibition committee, with a 2-line inscription *Légi Posta! Par Avion!* in a rectangle at the upper left. Figure 4-23 shows an example of this flown Budapest to Szeged. Mail, other than the cards and envelopes prepared by the committee, from these flights is quite scarce. Figure 4-24 shows a registered example of such mail flown on the Szeged-Budapest flight.

Mail exists in both directions bearing the vignettes with inscribed values, though their use was illegal. Ordinarily, postal markings do not tie the vignettes. This makes fakes quite easy to produce -- just soak off the vignette and replace it with one showing a printed value.

Also, there exists mail from both directions that bears the vignettes with values inscribed that were officially defaced. Postal officials did this by tearing or cutting-off the

Figure 4-24 Airmail Flown Szeged-Budapest

offending part of the vignette. These are known tied, which proves the defacement occurred prior to the flight. It is important to note that the defaced vignettes are identifiable as the first printing by their color. Tied examples of the vignettes with inscribed values, whether defaced or not, are both very scarce and interesting. Figure 4–25 pictures a postcard flown Szeged to Budapest that shows a tied, defaced vignette.

Figure 4–25 Airmail with Tied Officially Defaced Vignette

The **2nd Stamp Day** event was held in Budapest on December 12–14, 1925. A round-trip airmail flight between Budapest and Miskolc was scheduled. The flight did not take place. I believe the cause of the cancellation was the extensive flooding that occurred in eastern Hungary during December 1925. At least two airmail-oriented vignettes were prepared for the aborted flight. These are sometimes found on regular mail canceled with a roller cancel inscribed *II. Magyar Filatelista Nap*. One of these vignettes, printed in a brownish color with a black *Turul* carrying a letter, was later overprinted for use on the *4th Stamp Day* flight from Budapest to Györ.

An aerophilatelic commemoration did take place after the close of the *2nd Stamp Day* festivities. In honor of the 10th anniversary of the Przemysl balloon mail, souvenir cards were sent aloft on small model balloons. Latham states that 400 of these cards were flown. He is probably incorrect. Perhaps, 400 cards were carried *per* balloon, with several balloons released.

The light-weight cards are meant to look similar to the original Przemysl balloon cards. The *flimsy* card stock is in four colors -- green, buff, yellow, and blue -- but, the quality of the paper is poor and the colors have tended to fade. The cards were posted wherever the balloons were found, in early January. Exhibit officials prepared, and printed in four colors, a vignette for this mail. Usually only the dark green vignette is found on flown souvenir cards. Figure 4–26 shows a Przemysl 10th anniversary souvenir balloon card found and posted at Felsögod.

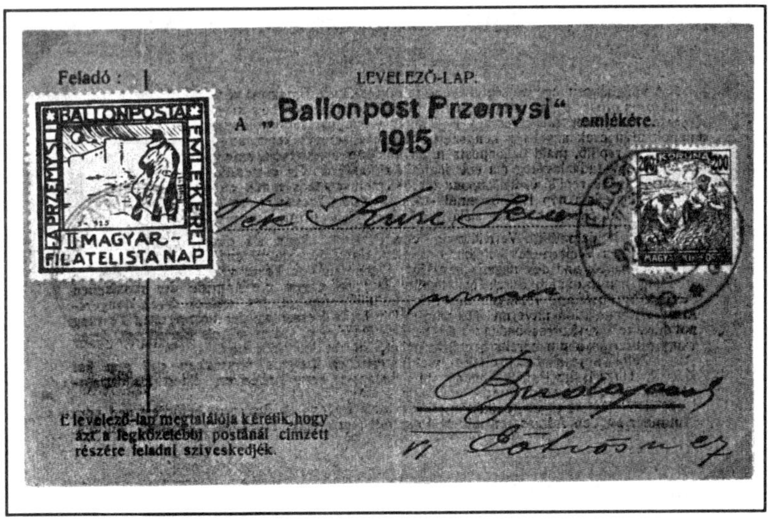

Figure 4–26 Przemysl Anniversary Balloon Card

Unfortunately, there was no airmail commemoration of the 1926 3rd Stamp Day.

During 1927, Hungary celebrated two special domestic airmail events. On April 17–18, a "Season Opening Flying Competition" was held at the Matyásföld airport. [At this time, many airlines still curtailed their flight schedules during the winter months.] This airmail, consisting of 1,000 numbered cards bearing a triangular red cachet, was flown over the field on the 17th. Figure 4–27 shows an example.

Figure 4–27 Season Inaugural Air Meeting of 1927

An airmail flight from Budapest to Györ was made on May 2, in conjunction with the **4th Stamp Day** celebration. Only specially printed cards (the reverse showing a Turul in flight) were carried. Markings on the front of each card include an *IV. Magyar Filatelista Nap* roller cancel, a lilac Budapest–Györ boxed biplane cachet, and a Györ circular date stamp receiving mark.

An overprinted vignette of the 2nd Stamp Day is often found on the reverse of the card.

Figure 4–28 Fourth Stamp Day Budapest–Györ Airmail

The number of such cards carried is unknown. Muller's catalog lists a Györ–Budapest return flight. No such airmail exists.

In 1928–32, several domestic air "routes" were initiated. These largely resulted from the expiration of most of the restrictions on domestic aviation imposed by the Treaty of Trianon. In the guise of these domestic air routes operated by MALÉRT, the beginnings of a Hungarian Air Force emerged. Generally, airmail is seen only from the first day of operation of these routes. While the routes continued to serve other purposes -- primarily military pilot training -- they were not very practical for speeding the mails.

On May 17, 1928 as in previous years, an air show was held at Mátyásföld; but, no airmail commemoration is known.

The first of the new domestic routes began in December of 1928 between Budapest and Szombathely. The airfield at Szombathely in an earlier time was, and in 1928 continued to be, a government–operated pilot training school. [Additionally, the government operated an aircraft mechanics school and an aviation research facility at Szekesfehérvár.]

The 1929 edition of Jane's mentions that trial flights were made between Budapest and Szombathely at the end of 1928 to test the airplanes produced by the *Manfred Weiss Aeroplane and Motor Works* at Csepel for airpost purposes. I have never seen airmail flown on this route; and suspect they were test flights, not mail carrying flights.

Latham mentions also a Budapest–Pécs–Kaposvár route and a Budapest–Nyiregyháza route of 1928. While these routes may have nominally begun operation in 1928, there is no real evidence of airmail from these routes until 1930 and 1931 respectively.

1929's contribution to Hungarian domestic airmail was the "reopening" of the Budapest–Szeged route. [Remember, this route existed in 1920–1 and then again for a single flight in 1925.] The airmail flown is probably limited to *the Second National Stamp Exhibition at Szeged* of May 9–12. Figure 4–29 shows the exhibit roller cancel on this mail.

Figure 4–29 Szeged–Budapest Airmail of 1929

In 1929, the Mátyásföld air show was held on May 19, again without an airmail commemoration.

International Airmail 1922–27: The Routes

With the end of World War I, the great wartime advances in aviation were applied to peaceful pursuits. One such pursuit was the development of commercial air services. To the post–war aerial entrepreneurs, airmail seemed to have great profit potential largely due to government subsidies.

Due to the chaotic situation in Hungary during 1919 and 1920, many other parts of Europe had access to airmail service before it was extended to Hungary. The nation's role in the growing European air network only began in 1922. Because of its central location, Budapest rapidly became a hub for European aviation in the 1920's.

Flown Hungarian mail of this period is actually much scarcer than one would expect. It is this author's belief that the low usage of airmail in Hungary resulted from a rate structure that was initially intended to discourage its use, for political reasons.

During the period that we are concerned with here, five airlines provided service to Hungary ... though, their names changed due to mergers. I list these airlines here and discuss them in more detail later.

Compagnie Franco Roumaine de Navigation Aérienne was known as CFRNA. In 1925, it merged to form CIDNA, and in 1933 it became a part of *Air France*. *Magyar Légiforgalmi Rt.*, usually called MALÉRT,

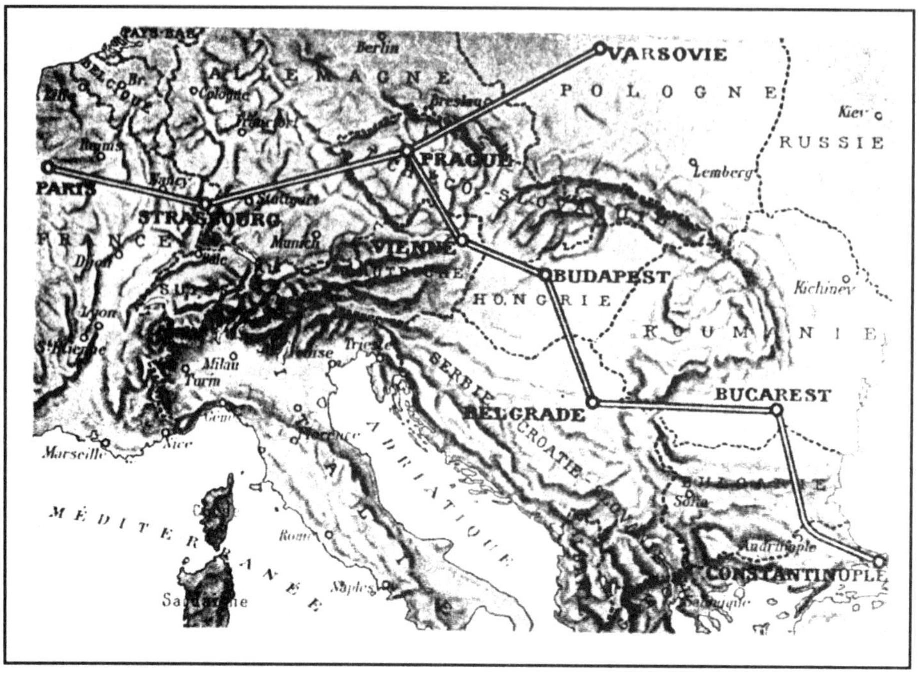

Figure 4-30 Map of 1924 CFRNA Airline Routes

was Hungary's privately owned, but government subsidized, national airline. ***Aeroexpress Rt.*** and ***Ad Astra*** were respectively the Hungarian and Swiss affiliates of the big German *Junkers* aviation company. They, with *Junkers*, became part of ***Deutsche Luft Hansa AG*** in 1926. Note: *Luft Hansa* was written as two words until after WWII. *Österreichischen Luftverkehrs AG*, known as ÖLAG was the Austrian *Junkers* company. It became the Austrian national airline in 1926. Until the Luft Hansa merger, the three *Junkers* companies operated jointly as *Trans-Europa Union* (TEU). Each of these airlines will be discussed in greater detail later.

Please note that during the entire period covered here, flying by all the airlines was restricted to the warmer months, essentially April through October.

CFRNA. On April 23, 1920 *Compagnie Franco-Roumaine de Navigation Aérienne* was established in Paris, to provide rapid transportation between France and the nations of the *Petite Entente* (Poland, Czechoslovakia, and Romania). Much of the capital to form the company came from Aristide Blank, who was president of the Romanian firm *Bank Marmorosh*.

Figure 4-31 1922 CFRNA *Potez IX* Biplane at Budapest Hangar

That year a Paris–Strasbourg route was established. In 1921 the route was extended to Warsaw via Prague. On May 1, 1922 a branch was added from Prague to Vienna and Budapest. In September 1922, CFRNA's routes finally reached Bucharest from Budapest. This route flew east via Arad, since the Yugoslavs would not allow access to any airplane coming from Hungary.

In 1923, the Yugoslavs relented and Belgrade replaced Arad as the stopover between Budapest and Bucharest. Figure 4–30 shows a map of the CFRNA route as it operated in 1923–24. [Survey and other occasional special flights were made from Bucharest to Constantinople; but, the Turkish port did not become a regular part of the route until 1926.]

As provisions of the Treaty of Versailles expired, Germany regained sovereign rights over its airspace. During 1924, Germany forced twelve CFRNA flights on the Strasbourg–Prague leg of its route to land, and seized the airplanes. They claimed these planes violated Treaty provisions because they met the specs for military aircraft being denied to Germany. Therefore, in 1925 the French airline (since merged and renamed to become *Compagnie Internationale de Navigation Aérienne* or CIDNA) altered its route to avoid German airspace.

The new route from Strasbourg to Vienna was via Basle, Zurich, and Innsbruck. This alpine route created many difficulties due to weather. The northeast branch of the CIDNA route flew from Vienna to Warsaw via Prague and Cracow. The diversion to Cracow was required so as not to overfly German eastern Silesia.

A Franco–German air agreement was reached in May 1926 and CIDNA returned to its original route with an additional stop at Nürnberg/Furth between Strasbourg and Prague. Also in 1926, CIDNA began regular service to Constantinople from Bucharest, and from Belgrade via Sofia.

In 1922, CFRNA was the only air carrier servicing Budapest. Figure 4–31 shows a CFRNA *Potez IX* plane at its Budapest hangar in 1922.

Throughout 1923 and 1924, CFRNA still predominated. By 1925, TEU was quite competitive on routes to the west, and provided the best connections to Germany and northern Europe through Munich. In 1926, with the merger that created *Luft Hansa*, that German airline took control of the Budapest market (except for the routes continuing eastward toward the Balkans, where CIDNA was the only choice).

I will now discuss and illustrate five covers flown from Budapest by CFRNA having special significance relating to the airline

Figure 4-32 May 1, 1922 "First Flight" Cover Budapest-Paris

itself and its routes. Other airmail carried by CFRNA-CIDNA is shown in the next two chapters.

CFRNA announced that airline operations to Budapest would begin on May 1, 1922. A handful of virtually identical Budapest covers bear a May 1 airmail cancellation, and Paris receiving marks of May 5. Whether or not they were actually flown, is a matter of opinion. Possibly more research will get a conclusive answer. Figure 4-32 shows an example of one of these controversial "first flight" covers.

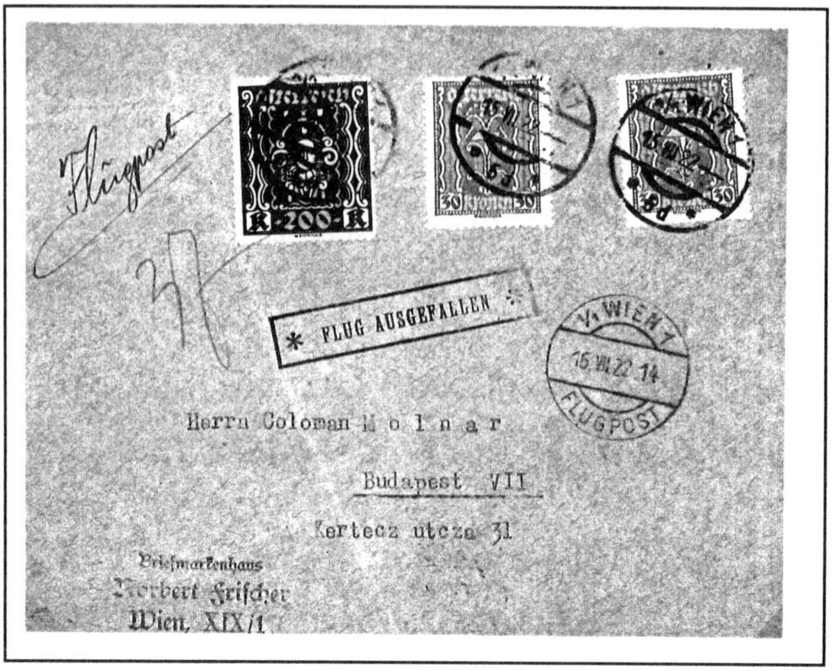

Figure 4-33 First Vienna-Budapest Airmail of 1922

Those concerned about the authenticity of these covers note that Hungary even lacked airmail regulations and rates until June 22 ... but, the May 1 covers are properly franked for the June 22 rates! Also, just because the airline was functioning doesn't mean airmail was flown. Finally, it has been said that this "airmail" could have gotten to Paris "on horseback in five days." Those who believe the May 1 airmail is authentic point out that the flight out of Budapest may have been delayed until May 4. Also, they note that it may have taken the postal authorities a while to "catch up" and publish the new regulations that had already been established. Despite the questions, many collectors consider these May 1 covers to be the first mail dispatched from Budapest by CFRNA.

During the summer of 1922, CFRNA flights to Budapest were very sporadic, since, even though it was temporarily the end of the route, Budapest had no maintenance facilities. Therefore, very little mail is known to or from Budapest before September 15 when the route was extended to Bucharest. There, a major new depot had been built for CFRNA operations to southeastern Europe.

On October 3, 1922, CFRNA announced the extension of its route from Bucharest to Constantinople. I believe that this was just a survey flight and a notice of intent. Normal air service to Turkey was not possible until the Treaty of Lausanne of 1924. Budapest-Constantinople airmail is not recorded until 1926.

Most authors attribute the first CFRNA mail-carrying flight from Budapest to dates in mid to late-July of 1922. It is well documented that the first Austrian mail (and probably the first of any origin) was flown from Vienna to Budapest by CFRNA on July 17. These covers, as illustrated in Figure 4-33, were postmarked on July 15 and also have a rectangular *Flug Ausgefallen* (translated "Flight Didn't Take Place") marking. The *Budapest 72* airmail receiving mark is of the 17th.

Two markings were in use on the airmail flown by CFRNA during 1922 and 1923. The most important is a pink rectangular "*Par Avion / Légi Pósta*" marking which was used

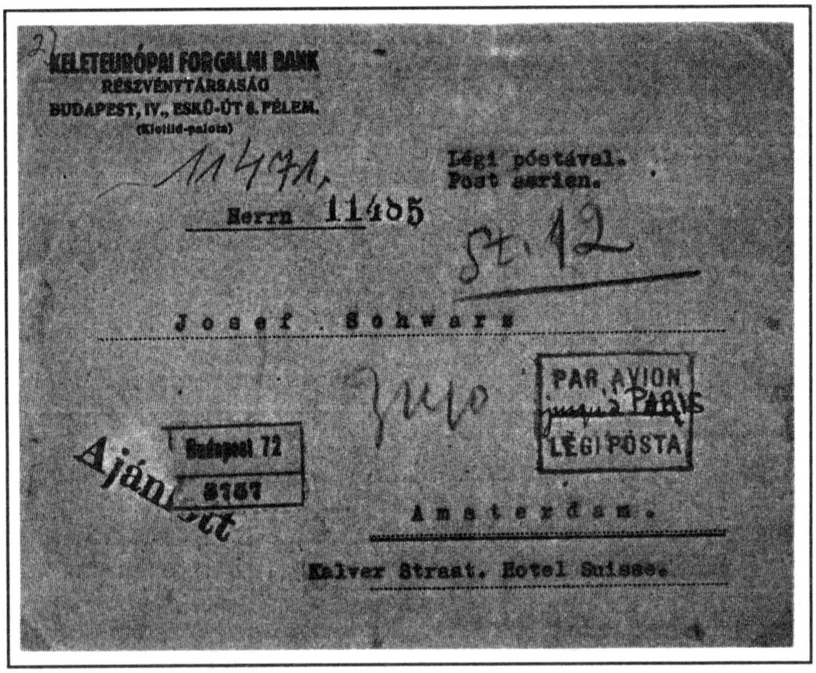

Figure 4–34 Budapest–Amsterdam *Jusqu'á Paris* Cover

on all but the earliest 1922–23 Hungarian airmail. It is also found occasionally on early 1924 airmail.

This marking served the same function as the airmail etiquette that came into use later. It simply highlighted a request for airmail service. It also sometimes served a secondary function -- as a "Jusqu'á" marking indicating how far toward the destination airmail service was expected. This was done with a manuscript notation on the horizontal line in the *Par Avion* handstamp. For example, Figure 4–34 has the notation "Jusqu'á Paris" indicating that this cover to Amsterdam was to be carried by air as far as Paris. On the reverse are 21 stamps totaling 64 Korona -- the correct rate for this *period 1a* date is 60K, the reason for over-franking is unknown. [Inflation *periods* are detailed in the next chapter.]

The second marking in use during 1922–23 is the rectangular (with concave corners) date handstamp of the CFRNA Budapest office. This black marking was not always used; actually, it is rather uncommon. It is usually found on the reverse of mail. Figure 4–35 pictures an August 9, 1923 cover flown from Budapest to Vienna with both of the 1923 markings. This *period 5* airmail is correctly franked on the reverse with 1,175K. Note that since the destination is on the airline route (i.e. Vienna) there was no need for a "Jusqu'á" notation.

Figure 4–36 shows a cover flown Budapest–Paris by CFRNA, and then on to London and Dublin. It is postmarked May 6, 1924 and

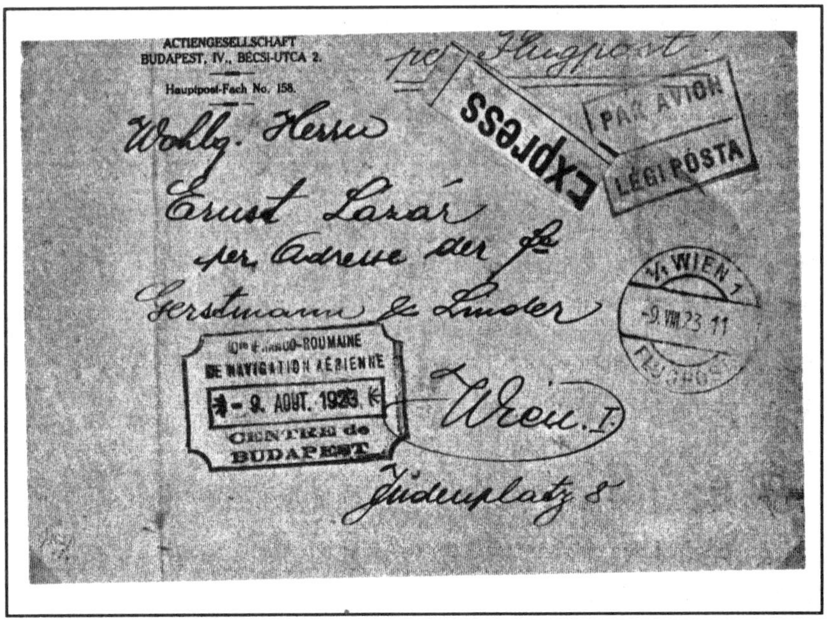

Figure 4–35 Budapest–Vienna Cover with CFRNA Marking

Figure 4–36 CFRNA 1924 *First Flight* to Paris

has additional franking of 20,000K on the reverse. This *period 10* airmail is overfranked by 100K, probably to show the low value of the 1924 airmail stamps.

May 6, 1924 is a day of special interest in Hungarian airmail. It was the date of CFRNA's first 1924 flights from Budapest. May 6, then, represents not only a "first flight" (of sorts), but also the first usage of the "boxed biplane" markings to Paris, Strasbourg, Warsaw, and Belgrade. [The Prague and Bucha–rest "boxed biplane" markings were not used initially in 1924, probably to make a political point.]

Magyar Légifor–galmi Rt. The terms of the Treaty of Trianon limiting use of aircraft in Hungary were lifted in November 1922, making possible the formation of Hungarian aviation and airline companies. The *Magyar Légiforgalmi Rt.* ("Hun–garian Air Traffic Company") –– generally called MALÉRT –– was formed on November 19. It began service between Vienna and Budapest on July 3, 1923. Though a private company, it received considerable government subsidies and became Hungary's *de facto* national airline. Figure 4–37 shows MALÉRT pilot, passen–gers, and *Fokker* plane at Budapest's airport.

During the 1920's, MALÉRT's international presence was limited to the Budapest–Vienna route. MALÉRT carried no mail in 1923; and, with a few exceptions, mail carried by them in subsequent years is not distinguishable. The reason is that CFRNA, MALÉRT, and TEU all operated between Budapest and Vienna and all carried some airmail.

Figure 4–37 MALÉRT Passengers Beside Fokker Airplane

Figure 4-38 MALÉRT First Flight of 1924

One MALÉRT flight to Vienna *has* distinguishable mail and is of particular interest to Hungarian airmail collectors. In 1924, MALÉRT made its first Budapest–Vienna flight of the season on May 2, a few days before the CFRNA first flights of the year. The result is "first flight" mail that also represents the first usage of the 1924 airmail stamps and of the "boxed biplane" marking.

This mail -- an example of which is shown in Figure 4-38 -- was postmarked either on April 30 or May 1. The Vienna receiving mark is of May 2. This *period 10* cover has 20,000K of additional postage on the reverse, so is overfranked by 600K.

Aeroexpress Rt. was formed on January 1, 1923 and began operations on July 15. It was a Hungarian company controlled by Germany's *Junkers* combine. Its Budapest to Geneva route flew via Zurich, Innsbruck, and Vienna. It was operated in conjunction with *Junkers* Swiss and Austrian affiliates under the name *Trans-Europa Union* (TEU). The route was flown by seaplanes landing on lakes or rivers near the various cities.

Very little airmail was carried by TEU in 1923. In later years airmail carried by TEU can sometimes be distinguished by the destination -- particularly Munich, which was their transfer point for routes within Germany.

Figure 4-39 shows a Junkers seaplane at the *Aeroexpress* facility on the Danube, near the Elisabeth Bridge. The first TEU airmail flight went from Vienna to Budapest on July 16, 1923. Figure 4-40 shows a cover carried on that flight.

Figure 4-39 *Aeroexpress* Seaplane at its Budapest Mooring

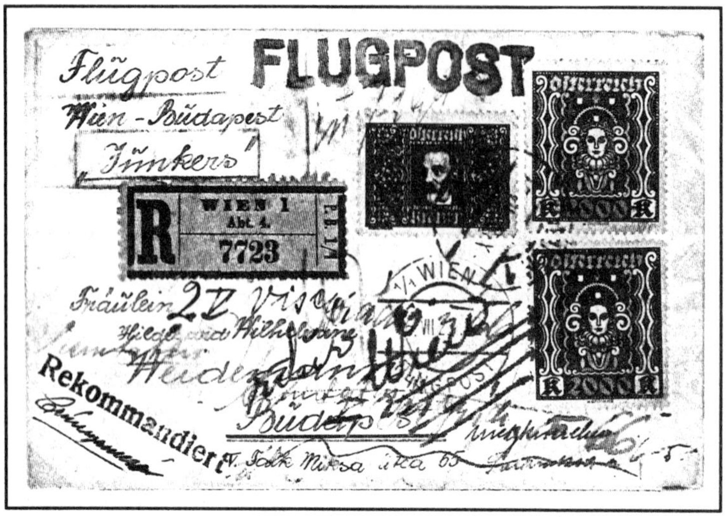

Figure 4–40 TEU Vienna-Budapest First Flight Airmail

From 1923–25, which of the three TEU airlines actually carried a specific piece of mail is always indeterminate.

Luft Hansa and ÖLAG. In 1926, when the *Luft Hansa* merger took place, the Austrian government successfully objected to the inclusion of *ÖLAG* in the merger. Therefore, from the perspective of the Hungarian aerophilatelist, *Aeroexpress* had two successors. They are *Deutsche Luft Hansa AG*, whose Budapest–Vienna–Munich route provided connections to Germany and northern Europe; and *Österreichischen Luftverkehrs AG*, whose Budapest to Vienna route connected to domestic Austrian services.

Figure 4–41 shows a typical Budapest–Munich cover flown by *Luft Hansa*. This *period 13* cover is correctly franked for the airmail printed matter rate.

Figure 4–42 shows a cover flown on the October 21, 1926 first flight of ÖLAG's new direct route from Budapest to Graz. It is overfranked by 1,000K for the *period 13* rate. Note that the Vienna "boxed biplane" marking was used. No such Budapest–Graz marking *ever* existed.

Other Flights. Two test or survey flight covers exist from the 1920's that are fairly scarce and little known. Both flights originated in Vienna and apparently made a stopover in Budapest. Please note that these do *not* represent official airmail; but, they are of peripheral interest to most Hungarian aerophilatelists.

Figure 4–41 ***Luft Hansa*** **Budapest–Munich Airmail of 1926**

The first was flown on a Vienna–Budapest–Vienna round trip in October 1923. It appears to be an ÖLAG test flight of a Fokker F-III airplane. This mail does not have a Budapest backstamp, indicating that it probably never saw a post-office within Hungary. Little other information about the flight is provided by this or other such mail I have seen. Figure 4–43 shows an example.

The company that made the second of these flights is unknown. The flight cards are post–

Figure 4–42 ÖLAG 1926 First Flight Budapest-Graz

marked in March 1925 and show two handstamps, as seen in Figure 4–44. The rectangular marking indicates the airplane's name was "Avis." The three–line marking says "Across Southeastern Europe" and then lists the cities the flight passed–through from Vienna to Sofia, Bulgaria.

These cards are only recorded with receiving marks of Pancevo, Yugoslavia and Sofia.

Figure 4–43 1923 ÖLAG Test Flight of F–III Aircraft

None has been recorded to date for the two Hungarian cities listed: Budapest and Új Palanka.

... the Rates (plus a 1922–23 Focus)

This chapter will cover three significant aero–philatelic concerns. First, the 1924–25 issue of airmail stamps that were in use from April 30, 1924 through March 31, 1927 will be discussed.

Then the detailed airmail postage rate structure during the 1922–27 period is reviewed. With the rate of inflation being most severe in 1922 and 1923, *and* airmail of those two years being quite uncommon, I will make a special point to illustrate that airmail. I will, of course, also show key pieces that illustrate the airmail rates of 1924–27, and some of the collecting options of that interesting period.

It is my belief that from 1922 to 1925 the Hungarian government for political reasons set airmail rates unrealistically high to discourage its use. This chapter will also explore that issue.

The 1924–25 Airmail Stamps. When CFRNA's international airmail service began in 1922, no airmail stamps were produced. The airmail stamps of 1920, issued for the domestic airmail service, were still technically valid. But, their denominations were too small for the rapidly escalating inflation. They can be found on a few

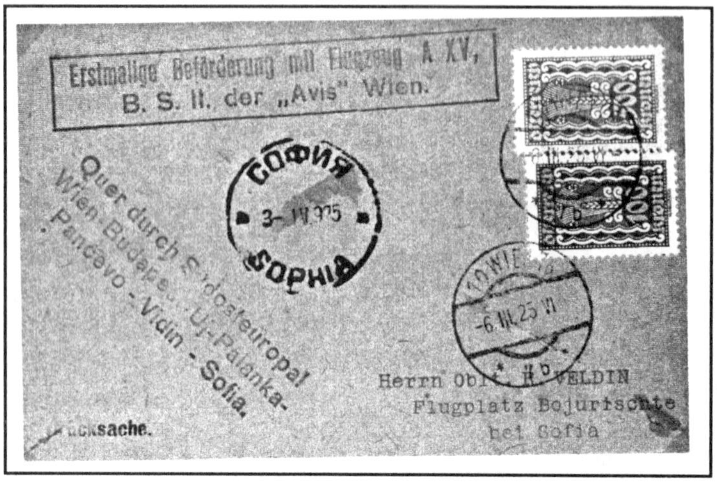

Figure 4-44 Flight "Across Southeastern Europe"

philatelic covers of 1922, in combination with higher-value regular stamps, but these are quite scarce.

Four airmail stamps were issued in 1924 for use at the inauguration of the flying season. These are the so-called *Icarus* issue in denominations of 100K, 500K, 1,000K and 2,000K. This issue was designed by Antal Megyer-Meyer. 415,000 perforated sets were printed on white paper with the "multiple double-cross watermark" in use since 1913.

Their first day of usage was April 30, 1924. Mail canceled on April 30 and May 1 was flown on MALÉRT's first flight of the season to Vienna on May 2. Figure 4-45 shows the low-value of this set.

Figure 4-45 The Airmail Issue of 1924-25

Obviously, the authorities were caught off guard concerning the rate of inflation over the winter of 1923-4. At the time of their release, the denominations of these stamps were altogether too small. This made for much colorful philatelic mail with a dozen or more airmail stamps often on a letter. It was not required that the airmail fees be paid by airmail stamps (unlike the 1918 and 1920 airmail services). Commercial airmail of 1924 is often franked with less colorful regular postage using higher value stamps. This commercial mail is comparatively scarce.

The situation was remedied in 1925 with the addition of two high values -- 5,000K and 10,000K. These two values were printed on paper with the newer "four double-crosses around a circle" watermark. The first day of usage of these stamps was April 20, 1925.

In 1926, with the end of inflation, reduction of airmail rates, and the elimination of the express (special delivery) requirement, these high-values became quite useless because they were too high-valued! Now they found an interesting internal post office use. Large blocks and sheets were canceled for accounting purposes. Basically this was to "balance the books" by canceling stamps to offset the cash received in payment for telegrams. These large used blocks, often on postoffice forms, are an interesting sidelight to an airmail collection.

Crude forgeries of the six stamps issued in 1924-25 exist. They are printed in sheets of 20 on very thin paper. The design is crude. The numerals indicating value are misplaced. The "watermark" is printed on the sheets with a waxy translucent ink. And, a Budapest "postmark" is also printed on the sheet -- but is a double-ring type never used as a Hungarian postmark. These forgeries of the 1924-5 Hungarian airmail stamps won't fool anyone, but they are scarcer today than the genuine unused stamps! Figure 4-46 pictures a full sheet of these forgeries.

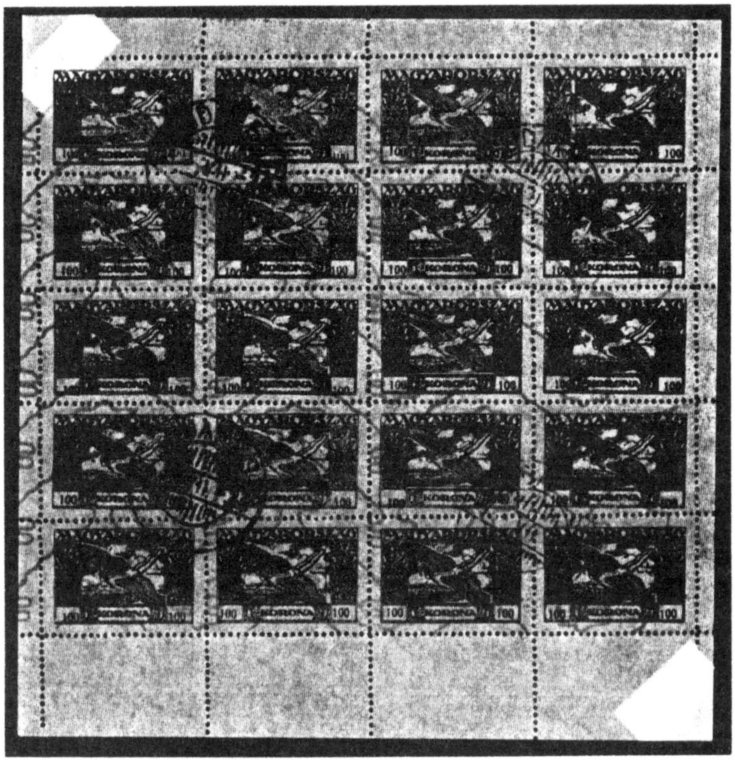

Figure 4–46 Forgeries of the 1924–25 Airmail Issue

The Airmail Rates. The inflation in Hungary from 1920 to 1925 was the result of several factors. Certainly the tumult of 1919, the disruption of transportation caused by the Trianon border changes, and the influx of ethnic Magyar refugees from the lost territories all played a major part. Our concern here is how that inflation affected airmail rates.

The table on this page provides the detailed data. The next several pages will attempt to explain and illustrate the information in that table. Superscripts are used to indicate notes which are explained later. To find an explanation of **data**[X], look for (**Note X**).

TABLE OF 1922–27 AIRMAIL RATES

(see notes in text for explanation of superscripts)

		AIRMAIL[1]		TREATY[2]		FOREIGN[3]			
Period	**Beg Date**	**Card**	**Lett**	**Card**	**Lett**	**Card**	**Lett**	**Express**	**Registry**
1	22Jun22	8K[4]	12K	3K	4.5K	4K	6K	12K	6K
1a[5]	01Aug22			5K	9K	7K	12K	24K	12K
2	15Sep22	14K	24K						
3	05Oct22	60K	100K	22K	40K	30K	50K	100K	50K
4	16Apr23	120K	200K	45K	75K	60K	100K	200K	100K
5	01Jul23	300K	500K	110K	175K	150K	250K	500K	250K
6	13Aug23	600K	1,000K	220K	350K	300K	500K	1,000K	500K
7	01Oct23	1,000K	1,600K	350K	600K	500K	800K	1,600K	800K
8[6]	01Jan24	1,600K	2,800K	600K	1,000K	800K	1,400K	2,800K	1,400K
9[6]	21Feb24	3,000K	5,000K	1,100K	1,800K	1,500K	2,500K	5,000K	2,500K
10	11Apr24	4,800K	8,000K	1,800K	3,000K	2,400K	4,000K	8,000K	4,000K
11	01Jun24	6,000K	10,000K	2,400K	4,000K	3,000K	5,000K	10,000K	5,000K
12	01Oct25	5,000K	10,000K	2,000K	4,000K	2,500K	5,000K	10,000K	6,000K
13[7]	17May26	2,500K	5,000K	16f	32f	20f	40f	80f	48f
14	01Jan27	20f	40f						50f

The regulations and rates for the 1922 airmail service were initially published on June 22. The required postage for a normal weight piece of airmail was **(Note 1)** the sum of the regular *Foreign* (or *Treaty*) postage rate, the *Airmail* surcharge, and the mandatory special delivery (i.e. *Express*) fee. The express fee was mandatory during periods 1–12. Often, mail did not have an "express" sticker or notation; but, special delivery *was* assumed for airmail.

Figure 4–48 Period 1 Letter Flown Budapest–Vienna

Based on bilateral postal treaties, **(Note 2)** reduced postage rates were available on all mail to Austria, Czechoslovakia, Germany, Poland, and Romania during the period under discussion. Some of the over–franking of early pieces appears to result from confusion about whether these bilateral postal treaties applied to airmail.

Basic letter rates **(Note 3)** for up to 20 grams are shown in the table, lower rates for printed matter and extra charges for heavy mail also applied; see Surányi's *Handbook* for details.

Figure 4–47 shows a postcard flown Budapest–Vienna on July 28. **(Note 4)** The airmail postcard rate went into effect on July 19, 1922. It is questionable whether *any* Hungarian airmail was flown by CFRNA before that date. This card has the correct 23K franking. Note that this piece does not have the rectangular *Par Avion / Légi Pósta* handstamp. It, and the May 1 cover shown in Figure 4–32, are the only 1922–23 Hungarian airmail items in my collection without it.

Figure 4–47 Period 1 Postcard Flown to Vienna

Figure 4–48 shows a *period 1* Budapest–Vienna airmail letter of July 29. Correct franking is 28½K. With the additional 10½K franking on the reverse, the total franking of this cover is 30K. It is overfranked by the amount of the treaty reduction.

(Note 5) Period 1a represents a change in the basic postal rates, without a change to the airmail surcharge. The Budapest–

Amsterdam cover shown in Figure 4–34 is an example of *period 1a*.

As the table shows, inflation affected postal rates greatly in 1922 and 1923. The foreign letter rate itself increased by over 100 times in only 16 months. *Period 2* reflects a change only in the airmail surcharge. Figure 4–49 shows a *period 2* airmail letter from Budapest to Warsaw, an unusual destination during this period. The correct franking for this letter is 57K. Its actual franking is 60K. Again, it is overfranked by the amount of the treaty reduction.

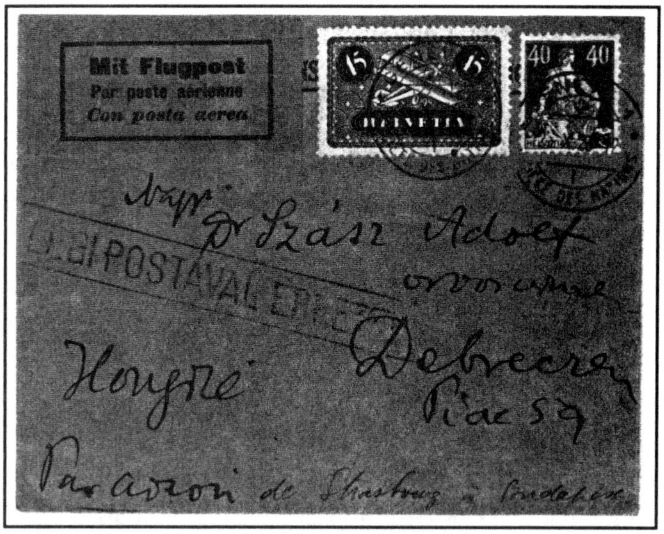

Figure 4–51 Geneva–Budapest Airmail of 1923

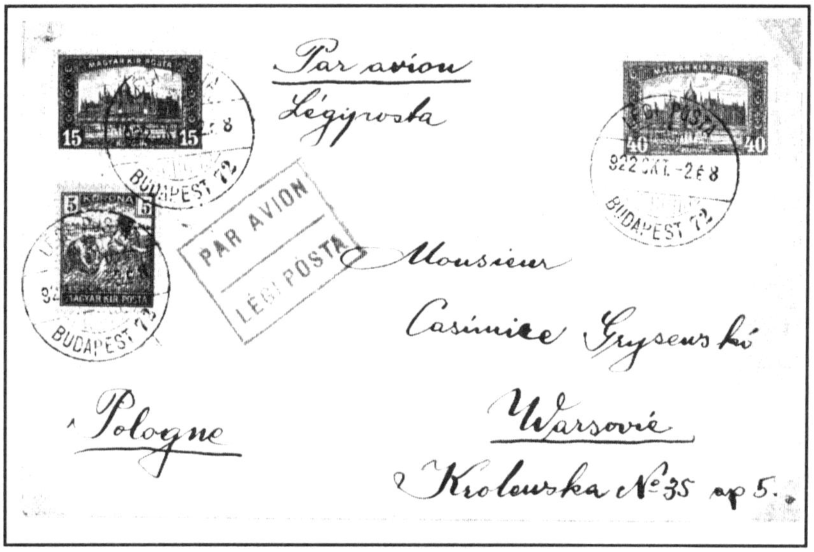

Figure 4–49 Period 2 Budapest–Warsaw Airmail

Figure 4–50 shows a *period 4* postcard flown Budapest to Belgrade on June 28, 1923, also an unusual destination. With a 381½K franking, it is overfranked only by the insignificant 1½K of the postcard itself.

The Budapest–Vienna letter shown in Figure 4–35 is from *period 5* in August 1923.

I also have enjoyed collecting airmail *to* Hungary. Virtually all of this mail for the 1922–23 period originated in Austria or Germany. Figure 4–51 shows an unusual letter flown from Geneva to Hungary. A manuscript notation indicates routing through Strasbourg, but this is not confirmed by any postal markings. This cover also shows the rectangular airmail arrival cancellation in use at Budapest during this period.

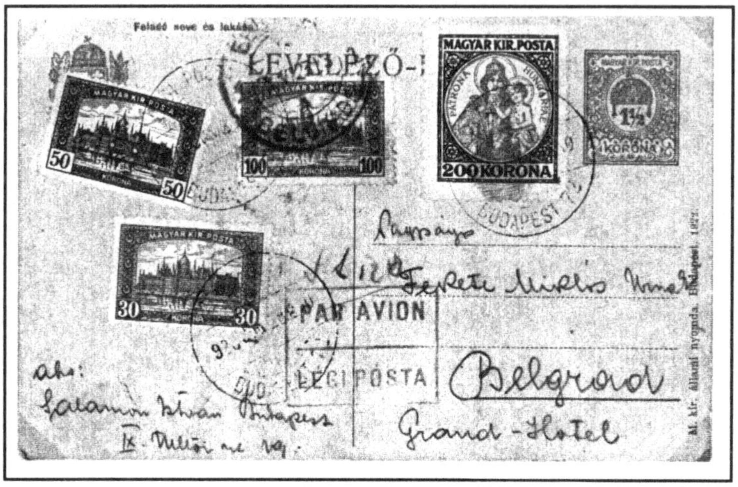

Figure 4–50 Period 4 Card Flown Budapest–Belgrade

The rate of inflation peaked in late 1923 and early 1924. **(Note 6)** As a result, three new airmail rates were enacted over the winter when no flying was done. So, there is *no* flown airmail of periods 8 and 9.

We have seen *period 10* airmail in Figures 4–36 and 4–38. It is the airmail from the early part of the 1924 flying season.

The bulk of the airmail of the mid–1920's is from *period 11*. This period began just before Hungary's inflationary spiral was stabilized by a loan from the League of Nations on July 2, 1924. It lasted sixteen months. *Period 11* airmail can be used to form an interesting study of the various esoteric postal rates of the time. For example, Figure 4–52 is a *period 11 registered printed matter* (!) cover from Budapest to Vienna. The postage for printed matter to Austria was 1,000K. The airmail surcharge for printed matter up to 150 grams was 6,000K. This cover is correctly franked at 22,000K.

The type of service requested makes it obvious this is a philatelic creation, not unlike much of Hungary's 1920's airmail. This fact is confirmed by the addressee. Sobetzky was Austria's foremost aerophilatelist. He was also an Austrian postal official. As we will see from time to time, he knew how to "make things happen" when it comes to creating airmail oddities.

In July 1925, a new Pengó currency was announced. It took a while for the postoffice to phase it in. **(Note 7)** Beginning on April 1, 1926, all postal rates (except the airmail surcharge) were stated in the new Pengó–fillér currency at the exchange rate of 1P = 12,500K. Postage stamps and postal stationery in the new currency had been issued on March 26; but, the Korona–denominated stamps continued to be valid until March 31, 1927.

This resulted in possible mixed frankings of stamps in the two currencies, during the entire 1926 flying season. Beginning May 17, 1926 with *period 13*, the airmail surcharge was halved and the requirement for mandatory express fees for airmail was dropped. Airmail became a lot *less* expensive.

Figure 4–53 shows a mixed franking on a postcard flown from Budapest to Vienna in November 1926. It combines the 8 fillér postcard of 1926 with 3,500K of airmail stamps of 1924. The total franking, which is correct, is 4,500K or 36f.

Another very unusual mixed franking, also of November 1926, is shown in Figure 4–54. It is an airmail cover addressed to the U.S. that was flown as far as Paris. At this date, it obviously "swam" across the Atlantic. This large registered letter was backstamped in New

Figure 4–52 Period 11 Registered Printed Matter Airmail

Figure 4–53 1926 Mixed Franking to Vienna

Figure 4–54 1926 Mixed Franking to New York

York, and is correctly franked at 16,000K or 1.20 Pengő. Note that it combines earlier regular *and* airmail stamps with 1926 regular stamps.

From April 1 to September 1, 1927, Hungary was without valid airmail stamps. During this time, a part of *period 14*, all airmail was franked with regular stamps denominated in the new currency. Figure 4–55 shows an example from August 1927 flown from Budapest to Prague. It is correctly franked with 72 fillér.

The airmail rates shown as *period 14* in the table persisted without significant change until World War II. This stability was the result largely of the worldwide depression of the early 1930's.

Was Airmail Discouraged?
In previous writings, I have discussed the idea that Hungary's government discouraged the use of airmail from 1922–26. I believe this was because the predominant airline serving Hungary was CFRNA, and it was largely controlled by Romanian interests. As the greatest beneficiary of the Treaty of Trianon and the cause of much misery during the turmoil of 1919 and 1920, Romania was the most hated enemy of Hungary and its people.

I believe there are two factors that prove this point. First, the real airmail costs to users were greatly reduced in 1926 when *Luft Hansa* took control of the airmail business in Budapest. Now, airmail could be encouraged because it was being carried by a "friendly" airline. Second, the comparable rate for airmail into Hungary was always about one–third of the rate for airmail out of Hungary.

The first part of my rationale is obvious from the table. The airmail surcharge reduction of *period 13* almost exactly coincided with the merger that created *Luft Hansa*. At the same

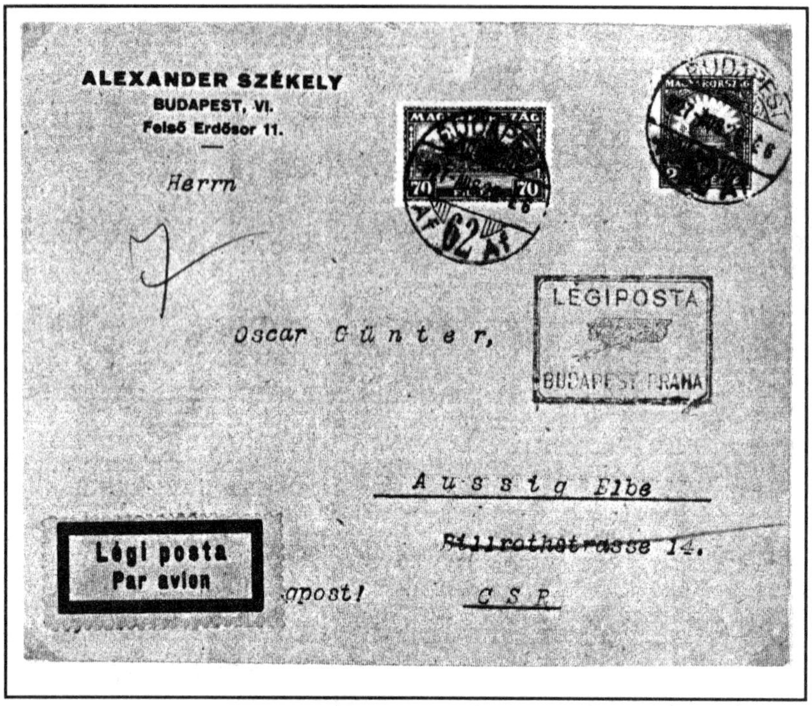

Figure 4–55 Budapest-Prague Airmail from August 1927

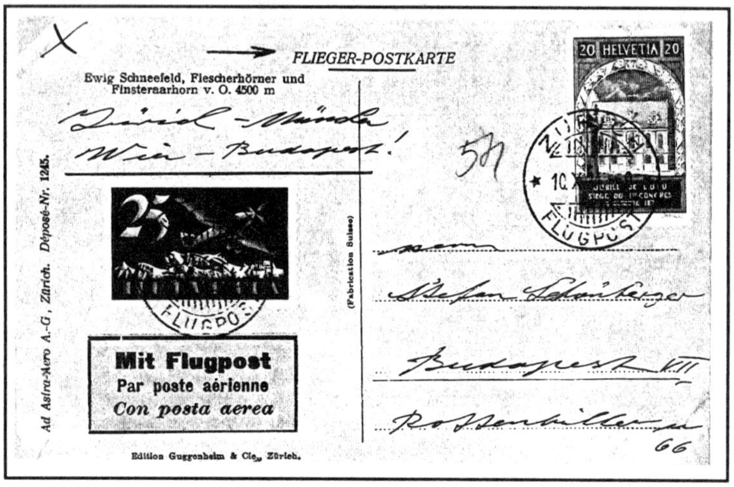

Figure 4–56 Example of Low Swiss Airmail Rates

time, the *express* fee was no longer mandated. This resulted in a 70% reduction in the actual cost of airmail to users.

The second point I will try to prove by comparing actual airmail rates between two cities. As mentioned previously, the currency of Hungary was stabilized in 1924 by a loan from the League of Nations. In July 1925, the new Pengő currency was introduced. During the intervening year, postage rates were stable, and so was the exchange rate -- at about 14,000K = 1 Swiss Franc.

Figure 4–56 shows an October 1924 Swiss airmail postcard from Zurich to Budapest. It cost its sender 0.45 SFr. A postcard from Hungary to Switzerland at the same time would have cost its sender 19,000K or 1.36 SFr -- three times as much. See *period 11* in the table of airmail rates for the detailed breakdown. This 3:1 ratio had been approx–imately the case since the inception of the airmail service in 1922.

Hungary maintained this 3:1 cost ratio by a high airmail surcharge and mandating the use of special delivery on airmail. About this same ratio applied to all forms of airmail. For instance, in Figure 4–57 you see an airmail letter from Budapest to Zurich. It is from the same period -- September 1924. The 17,000K with which it is franked (an additional 13,400K is on the reverse) is the correct franking for the printed matter rate. This is equivalent to over 1.2 Swiss Francs. I hope the comparison of these two airmail items proves my thesis.

Of course, this analysis doesn't change anything about how we collect. It simply describes an unusual twist of history. Perhaps it is important, though, because it tells us why there was so little real commercial airmail originating in Hungary before 1926.

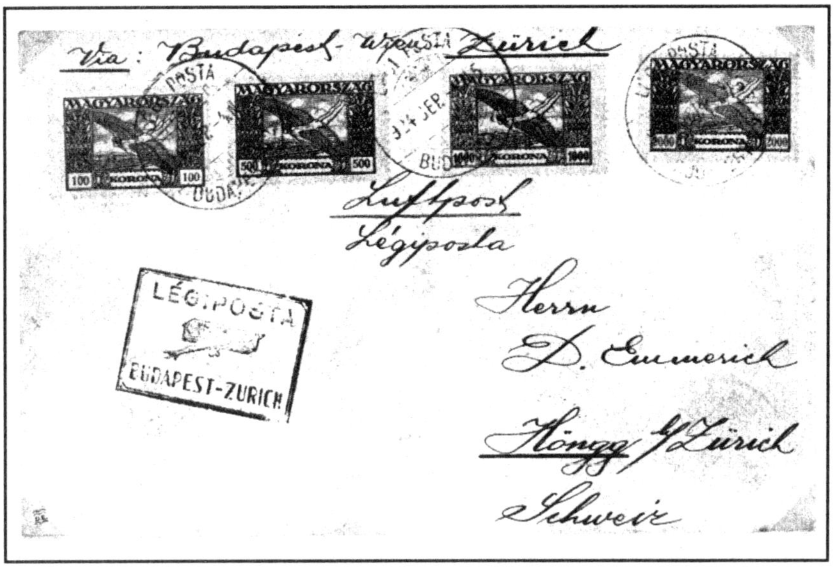

Figure 4–57 Example of High Hungarian Airmail Rates

... and the "Boxed Biplane" Cachet

The most notable feature of Hungarian airmail of the 1924–27 period is the "boxed biplane" route marking. This cachet was originally prepared for use between Budapest and Vienna, Strasbourg, Paris, Belgrade, and Warsaw for the opening of the 1924 flying season. Each indicated an airmail destination on a direct route from Budapest. Between 1924 and 1927, this marking was used on airmail for 18 foreign destinations.

"Proofs" of this marking to other cities exist, but these never appear to have been used. Five "boxed biplane" domestic markings were also prepared and used on special philatelic exhibition flights, as noted in the chapter on the *Domestic Airmail Events of 1924–29*.

Figure 4–58 "Boxed Biplane" Proofs

Proofs of the original five markings exist with and without clouds around the biplane. The clouds obscured the image, so the cachets put into use had no clouds. Figure 4–58 shows an example of the proofs in both forms. Note that the unissued "Strassbourg" cachet erred by combining the French and German spellings of that city's name.

The five cities initially chosen for the markings are all on the CFRNA routes. Vienna, of course, was also on the MALÉRT and TEU routes. It is interesting to conjecture why the two remaining CFRNA destinations, Bucharest and Prague, were not initially included. I suspect it was political. Romania and Czechoslovakia were the most despised of the successor states.

Boxed Biplane Cachets on Airmail

City	Color	ERU	RR
Wien	Black	1924	2
Beograd	Black	1924	5
Paris	Black	1924	2
Strasbourg	Black	1924	3
Warszawa	Black	1924	4
Bucharest	Black	1924	6
Zürich	Black	1924	5
München	Black	1925	3
Berlin	Black	1926	5
Constantinople	Black	1926	7
Köln	Black	1926	6
Praha	Black	1927	4
Dresden	Black	1927	3
Wien (Sep27)	Black	1927	1
Innsbruck	Black	1927	3
Breslau	Black	1927	2
Frankfurt/Main	Black	1927	2
Leipzig	Black	1927	2
Nürnberg	Black	1927	2
Wien	Red	1927	3
Dresden	Red	1927	8
Paris	Red	1927	9
Praha	Red	1927	9
München	Red	1927	10

The table above shows all the markings that were used on mail; and, it identifies the color, the earliest recorded usage (ERU), and a rarity factor (RR) for each.

Regarding ERU, I would not be surprised if earlier usages are found for some markings. The earliest recorded red marking is June 1927. Also, I wouldn't be surprised to find red 1927 markings for additional cities. I would be quite surprised, though, if boxed biplane cachets of cities not in the above table are found on genuinely flown airmail.

The scale used for RR is this: the most common boxed biplane marking (late usage of the black Budapest-Wien) is arbitrarily given a 1; the scarcest (the red Budapest-München) arbitrarily given a 10. The rest are placed in between. When the new airmail stamps were issued in September 1927, large numbers of obviously philatelic covers were flown bearing all the boxed biplane markings. These are much more common than earlier usages. The rarity factors in the table, for markings which existed before September 1927, are for earlier covers. The last five black markings on the list did not exist earlier, so obviously can only be found used in September 1927 or later.

It's important to realize what the "boxed biplane" marking is, and what it is not. It is *not* a *Jusqu'á* marking. Those markings indicate the point at which airmail service is to end. While some misunderstand and believe that is the case, you will see examples that completely contradict their view. For instance, Figure 4-66 shows a cover flown from Budapest to Copenhagen. The cover itself gives evidence of that fact. Yet the marking is *Budapest-Strasbourg*.

The boxed biplane cachet is simply a route director marking. It shows on which direct route from Budapest the mail was carried. The cover noted above was carried by CFRNA to Strasbourg. From there, it went via KLM to Amsterdam, Hamburg, and Copenhagen. This same idea is shown over-and-over again by Hungarian covers to England and northern Europe.

Knowing what the markings look like, which varieties were used on airmail, and what function they served, let's examine some examples.

Figure 4-59 Budapest-Wien Boxed Biplane Cachet of 1924

Figure 4–61 October 1924 Bucharest–Budapest Airmail

The table gives some idea of the frequency of usage of airmail to the listed cities. Vienna is the most common as would be expected. The scarcity of airmail from *and* to Hungary varies considerably from country to country. In general, the reasons for the differences in volume are obvious.

The close business relationships with Austria survived the breakup of the Dual Monarchy. Austrian and German commercial airmail to Hungary is quite common. Both philatelic and commercial airmail from Hungary to Austria (and by 1925 also to Germany) was also quite common.

Airmail to and from Hungary's successor states -- Romania, Czechoslovakia, and Yugoslavia -- is quite scarce as would be expected. These countries were not on friendly terms with Hungary. Polish, British and Swiss mail is less scarce. Airmail from the far–flung corners of Europe, such as the Scandinavian countries, is also occasionally found.

Airmail to France from Hungary is not uncommon, and this includes a good deal of very nice commercial mail. Most surprising, though, is the apparent great scarcity of airmail from France *to* Hungary.

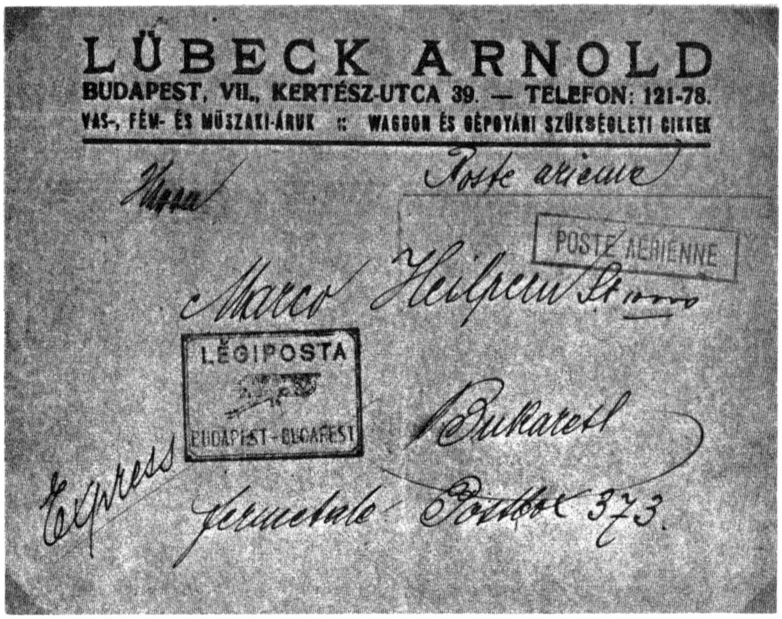

Figure 4–60 Budapest- Bucharest Cover of November 1924

Figure 4–59 shows a large June 1924 cover flown Budapest–Vienna. The franking with a large number of airmail stamps is very pretty; but, it is notable that it is only slightly over–franked. Clearly, in June 1924 already the denominations of the airmail stamps were insufficient to meet inflationary needs. Most airmail covers, unlike this one, used high–value regular stamps to supplement a few

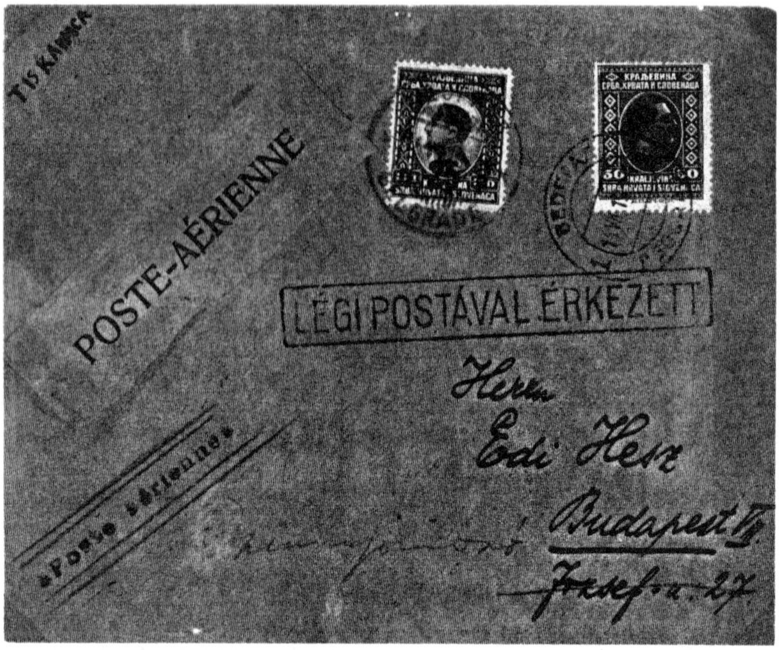

Figure 4–62 Airmail Flown Belgrade–Budapest

airmail stamps. Please note also that this cover has the 1922–3 rectangular *Par Avion / Légi Pósta* marking in addition to the boxed biplane cachet. This marking is seen only occasionally on early 1924 mail.

Figure 4–60 and 4–61 show airmail flown to and from Bucharest late in the 1924 flying season. The cover in Figure 4–60 is franked with 24,000K in regular stamps on the reverse, and also bears a Bucharest receiving mark. Romanian airmail specialists consider these two covers among the greatest rarities of Romanian aerophilately.

Figure 4–62 shows a 1925 airmail letter from Belgrade to Budapest. It is noteworthy that, though Yugoslavia had no airmail stamps until a decade later, a special cancellation was used on the postage representing the airmail surcharge. This is not apparent from the figure, because the difference was the color of the ink. The stamp at the right -- the airmail surcharge -- is canceled in red. This item has a Budapest receiving mark of July 17.

Figure 4–63 shows a June 1924 commercial cover from Budapest to Paris. Many collectors find such commercial airmail more "collectable" than the gaudy philatelic alternatives. On the reverse of this piece of airmail is additional franking of 20,000K in regular stamps and a Paris receiving mark.

Note that, in addition to the boxed biplane cachet, it also has a pink rectangular *Poste Aérienne* handstamp. I believe this was a CFRNA marking that superseded the 1922–3 rectangular marking. It is found only on 1924

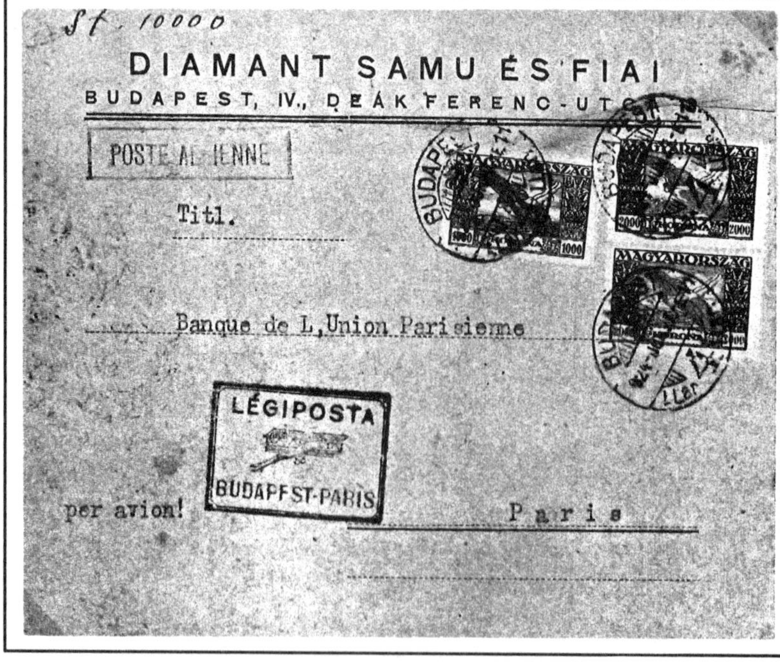

Figure 4–63 Commercial Budapest–Paris Airmail

Figure 4–64 Paris–Budapest Airmail

airmail, and usually only on commercial mail to Paris. There are exceptions to this last "rule." You may have noted that the Budapest–Bucharest cover in Figure 4–60 also has this handstamp.

Figure 4–64 is a 1925 postcard flown from Paris to Budapest. It is the only piece of French airmail to Hungary from the 1922–27 period that I have seen in my thirty years of seriously collecting Hungarian airmails.

The October 1925 cover seen in Figure 4–65 also has a Budapest–Paris boxed biplane marking. With a picture of a Zeppelin and an inscription that says "Airmail Budapest–London, England," it turns many heads. The hidden part of the inscription is even more interesting. It reads: "with Connecting Dispatch Airmail from London to New York on the First Trip by the U.S. Airship Los Angeles." This cover is not particularly uncommon, and I believe its creator had honest intentions. When it became apparent the mail would not be carried by the *Los Angeles*, he pasted the stamps over that part of the printed inscription.

Germany turned the Zeppelin *LZ–126* over to the U.S. in 1924 as a part of war reparations. On several of its early flights, that airship -- renamed the *Los Angeles* -- carried commercial airmail. This practice ended in the summer of 1925, before its London–New York flight. So this cover flew from Budapest to Paris, and maybe as far as London, but it certainly traveled by sea across the Atlantic.

Figure 4–65 Airmail Intended for Zeppelin *Los Angeles* Flight

Airmail between Hungary and Scandinavian destinations was very uncommon. Figure 4–66 shows an October 1924 airmail cover from Budapest to Copenhagen. With an additional 20,000K of regular stamps on the reverse, it is philatelicly overfranked by the value of the airmail stamps. It was flown from Budapest to Strasbourg by CFRNA. From there it went to its destination via Hamburg, Germany.

In Figure 4–67 you see a July 1925 postcard from Sweden to Hungary via Berlin

Figure 4–66 Budapest–Copenhagen Airmail

and Dresden. Note that the addressee is *Julius Pewny*. A significant amount of Budapest's philatelic airmail of the period was either sent by, or to him. He was the manager of the Budapest office of CFRNA/CIDNA.

Many very special boxed biplane airmails are from 1926. It was the year mixed frankings were possible, as seen in Figures 4–53 and 4–54. Also, a couple of new destinations emerged with boxed biplane cachets.

Figure 4–68 shows a very unusual September 1926 cover from Budapest to the Rheinland, via Munich and Cologne. It is an example of one of the new *Luft Hansa* routes, now that virtually all the German airlines were integrated. The *Budapest–Köln* boxed biplane marking was rarely used that early.

This cover also has two other distinctions. It is a mixed franking. It is franked on the reverse by a 10,000K Madonna stamp and 72 fillér in regular stamps of the 1926 issue. This represents the correct franking for *period 13* airmail special delivery -- 1.52 Pengó. Remember, by *period 13*, the special delivery (express) fee was no longer mandatory. Therefore, special delivery was rarely used.

Though occasional flights were made by CFRNA/ CIDNA to Constantinople earlier, the former imperial city of the Ottoman Empire was not a routine destination until 1926. This is also when we first see usage of a *Budapest–Constantinople* boxed biplane cachet.

Figure 4–69 shows a postcard flown Budapest–Constantinople in June of 1926. It is philatelicly overfranked. Figure 4–70 is an October 1926 Turkish cover flown from Constantinople to Budapest. Both are quite unusual.

Figure 4–67 Sweden–Budapest Airmail of 1925

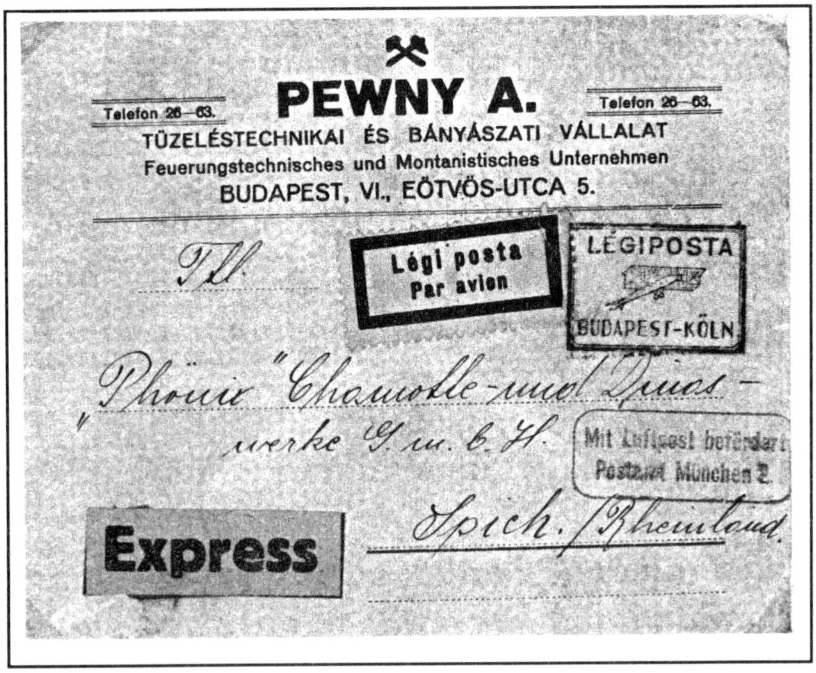

Figure 4–68 Budapest–Köln Boxed Biplane Marking

Transit Markings. The boxed biplane marking was intended, and was almost exclusively used, as a *route director* marking applied at the origination point of the mail (Budapest). It was also very occasionally used as a transit marking, on German airmail passing through Budapest toward the Balkans. The three known transit markings are: (1) mail from Mannheim via Budapest to Bucharest with the *Budapest–Bucharest* marking; (2) mail from Bonn via Budapest to Sofia with the *Budapest–Beograd* marking; and (3) mail from Munich via Budapest to Constantinople with a *Budapest–Constantinople* boxed biplane marking. Figure 4–71 shows an example of the latter.

Since each of these routes involved a transfer from *Luft Hansa* to *CIDNA*, it seems probable that the boxed biplane cachet in these cases was applied by CIDNA personnel.

Late 1927 Markings. The validity of the airmail stamps of 1924–5 ended on March 31, 1927. On September 1, 1927 Hungary issued airmail stamps in the new Pengő–fillér currency. The usage of these 1927 airmail stamps for the next several years is discussed in the chapter on *International Airmail of 1927–29*.

Because several of the boxed biplane route markings are known only from late 1927, and therefore on mail with the new stamps, that aspect of 1927 airmail will be covered here.

Figure 4–69 Budapest–Constantinople Airmail

The five destinations that are only recorded from September 1927 are Innsbruck, Austria and four German cities -- Nürnberg, Leipzig, Breslau, and Frankfurt am Main. These now all had regular connecting service from Budapest as a result of the DLH consolidation.

At this point, some comment about the value of much of

Figure 4-70 Constantinople-Budapest Airmail of 1926

the boxed biplane mail of late 1927 may be in order.

From late 1927 (mostly September) large numbers of not very interesting looking pieces of airmail exist, with all of the eighteen boxed biplane destinations represented. This mail is worth much less than earlier mail with the same markings.

Figure 4-71 Hungarian Transit Mark on German Airmail

The boxed biplane cachets of cities such as Belgrade, Bucharest, Constantinople, and Cologne of earlier years are significant rarities. Those of late 1927 are not cheap; but, they are certainly in a different "class."

The comments above obviously don't apply to mail of Innsbruck and the four German cities listed, since boxed biplane mail to them only exists from late 1927. Therefore, these are more sought after than other covers of the same period. Figure 4-72 shows the Budapest-Innsbruck boxed biplane cachet on an envelope correctly franked with 26 fillér for the printed matter rate.

Red Markings. Surányi's *Handbook* only reports the boxed biplane marking in red for *Budapest-Wien* and *Budapest-Dresden*. A few years ago, I reported a *Budapest-Paris* example, and at least one other collector has reported the same.

Recently, I came upon red boxed biplane markings for *Budapest-Praha* and for *Budapest-München*. To my knowledge those cities have not been reported by anyone else.

What else is known about the red markings? They all are on airmail postmarked in 1927. The earliest, my Budapest-Paris, is dated June 22. The latest is the Budapest-München which at November 12 is the latest date I've seen any boxed

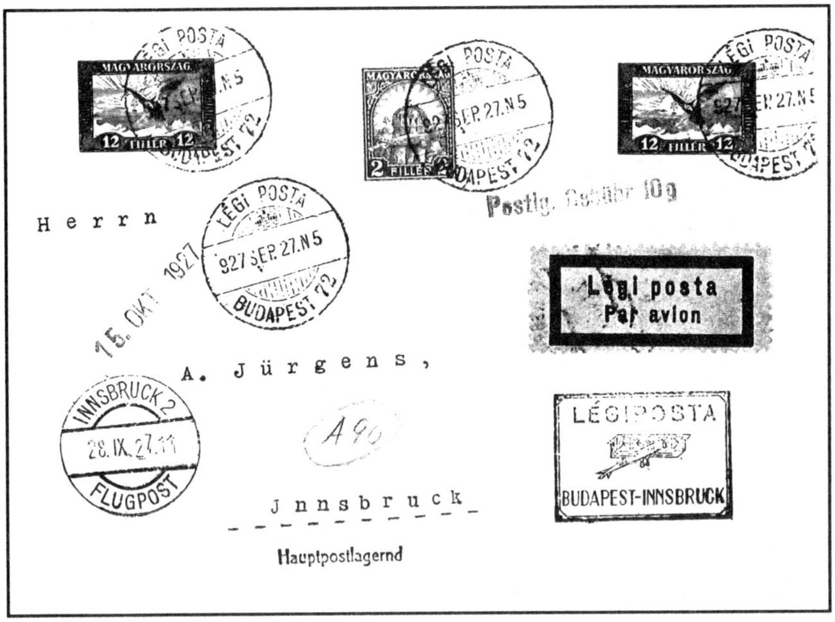

Figure 4-72 Budapest-Innsbruck Boxed Biplane Marking

biplane cachet used. Four of my five red boxed biplane markings are on what appear to be commercial covers -- though two of those are from stamp dealers. Will additional red boxed biplane cachets be found? That seems likely, but almost certainly they will be dated 1927.

I have chosen not to illustrate any of the red boxed biplane covers, since on these printed pages they would be indistinguishable from those applied in black ink.

Other Cities. Boxed biplane markings of other cities do exist, but *not* on mail. These canceled-to-order pieces are usually found on a plain square of paper ... less frequently tying cheap Hungarian stamps to a piece of paper. Please note: the legitimate boxed biplane cachets were *never* used to cancel the stamps on mail. How did these occur? Possibly someone had the opportunity to play with a handstamp and was clever about changing the type.

Alföldi and Talas report these markings: *Angora* (Ankara), *Bale* (Basle), *Danzig*, *Kraków*, *Lwów*, *Genève*, and *Lausanne*. I think I recall also seeing in the past *Sofia* and *Genf* (the German spelling of Geneva). All of these cities either had direct (or relatively good) connections from Budapest. So, it is also possible that the hand-stamps were officially prepared for future use, which never occurred. Still other cities may exist. In any case, they were not used on mail.

One other boxed biplane marking is claimed to have been used on actual airmail. This is *Budapest-Graz*. According to both the Sobetzky and Kosel catalogs, two examples of this marking exist on covers.

I had the privilege of seeing one of those. It is a postcard flown on the April 23, 1927 first Budapest-Graz flight of the year. The *Budapest-Wien* boxed biplane marking was lightly applied and then the word *Graz* was written with pen-and-ink over the word *Wien*. This card was addressed to the distinguished Austrian postal inspector -- Georg Sobetzky! The message on the card says something to the effect "As you requested, I'm sending you two of these cards, one registered and one normal mail." I have no idea whether the Graz "pen marking" was added before or after the flight. My assessment of the Budapest-Graz boxed biplane marking is the same in either case. I wish someone hadn't messed up a nice piece of airmail!

In late 1927, the use of the boxed biplane route marking ended. In some ways it also signaled the end of *The Pioneer Period of Hungarian Airmail*. Aircraft were improved, giant airlines such as *Deutsche Luft Hansa* had consolidated, airline routes criss-crossed all of Europe, and for the first time airline service was year-round. At least, that was true west of Budapest. There were still problems with winter flying in the Balkans.

International Airmail of 1927-29

In 1927, there were still only three airline routes out of Budapest, and four airline companies operating them. CIDNA operated the route to Belgrade. ÖLAG operated the route to Graz. All four -- *CIDNA*, *MALÉRT*, *DLH*, and *ÖLAG* -- operated on the Budapest-Vienna route. Other routes were authorized; for instance, MALÉRT had authority to operate routes to Prague, Belgrade, and Zagreb. These were never actually flown, probably because of the on-going dis-agreements between Hungary and the Czech and Yugoslav governments. It was 1929 before another airline came to Budapest.

Figure 4-73 The Airmail Stamps of 1927 on Cover

The Airmail Issue of 1927-30. September 1, 1927 saw the release of eight airmail stamps in the new Pengő-fillér currency. Figure 4-73 shows these stamps on an airmail cover to Germany.

Four values were added to the set on October 16, 1930. The entire set was valid until the end of 1933. Figure 4-74 shows the 1930 values on an early airmail cover to England.

The combined issue of 1927-30 included eight stamps picturing a *Turul* and a sunrise.

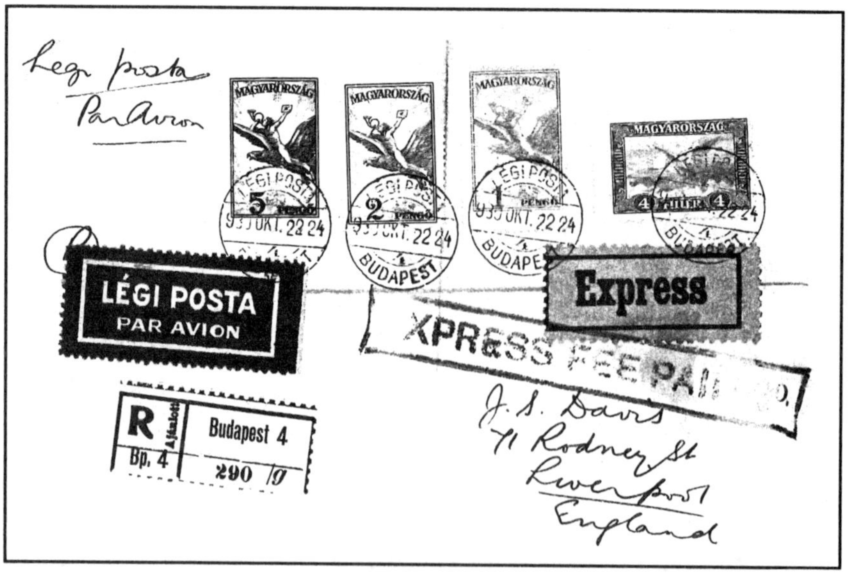

Figure 4-74 Airmail Stamps of 1930 on Cover to England

Figure 4–75 The Airmail Issue of 1927–30

János Pálinkás designed these stamps. Figure 4–75 shows the 40 fillér value of this design.

Jenö Haranghy designed the four stamps picturing Mercury with a post horn atop a *Turul*. Figure 4–76 shows the 5 Pengó value of that design. Quantities of the airmail issue of 1927–30 ranged from 110,000 to 314,000. These stamps are found imperforate.

New European Connections. European air connections improved greatly as new aircraft permitted longer non–stop flights, carried more paying passengers, and were less susceptible to weather cancellations.

One important new destination was the Soviet Union. By 1926, a joint venture between *Deutsche Luft Hansa* and the *Russian State Airlines* (later *Aeroflot*) was intent upon establishing a route between Berlin and Moscow. Figure 4–77 shows an airmail letter flown Budapest–Berlin, and then on to Smolensk on that Berlin–Moscow route. Smolensk, now on the westernmost edge of the new Russian Republic, was one of the two stopovers on this route.

Figure 4–77 Budapest–Berlin–Smolensk Airmail

Figure 4–78 is a 1929 postcard flown from Budapest to Copenhagen. It is unusual in that it is a Danish reply card with additional Hungarian franking to pay the airmail surcharge of 20f plus the difference in regular postal rates. Note that the 32f Hungarian stamp does *not* pay the full airmail postage of 40 fillér.

This card is also interesting because the inscription gives one an idea of the routings and timing considerations for airmail in

Figure 4–76 Airmail Issue of 1927–30

Figure 4–78 Budapest–Copenhagen Airmail of 1929

Europe, where direct routes didn't exist. As indicated in the inscription, the flight time was 12 hours and 10 minutes. Both the Budapest postmark and the Copenhagen receiving mark are of August 14. Yet, the card made stops at Vienna, Prague, Chemnitz, Halle, Hannover, and Hamburg along the way!

New Airlines in Budapest. Two additional airlines began service through Budapest in 1929, *KLM* and *Imperial Airways*. For both, Budapest was a stop on a long route to Asia.

On September 12, *KLM* began its bi–weekly Amsterdam–Batavia service to the East Indies. Until it ceased operation in 1939, this was the longest scheduled airline service in the world. Initially, using Fokker F-VII airplanes, 18 stopovers were made on a 12–day odyssey totaling 89 flying hours. European stops were at Nürnberg, Budapest, Sofia, and Istanbul.

Latham mentions Hungarian airmail carried from Budapest in both directions on the Amsterdam–Batavia route. His entries include first–flight mail and other mail in the autumn of 1929. I have never seen examples of such airmail flown during this first year of operation.

In the 1920's and 1930's, Britain's *Imperial Airways* was primarily interested in setting up airline service to the British colonies in Asia and Africa. In Europe, excepting the lucrative London–Paris route, *Imperial Airways* was almost non–existent.

The natural starting point toward both the Asian and African colonies was Cairo. Due to the poor European service offered by *IA*, special arrangements were required to get airmail and air passengers there from London. Several combinations of planes, boats, and trains were tried. The first all air service was provided, via Budapest, in November 1929.

This route lasted only two flights. The flying conditions over the Balkans were too poor for the equipment of that day. In April of 1930, the air route was resumed, excepting the short segment from Skopje, Yugoslavia to Salonika, Greece. This was made by train. This routing continued until May 1931 when *IA* went back to train service from Basle to Brindisi, Italy. Henceforth, Budapest was not a part of the *Imperial Airways* route to Asia.

Mail acceptance for Budapest was only available on the second flight of 1929. Figure 4–79 shows a rare cover flown London–Budapest on that flight of November 9.

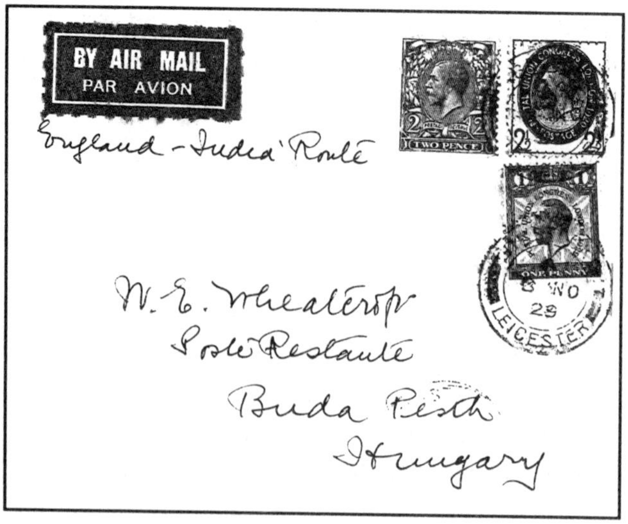

Figure 4–79 London–Budapest 1929 Airmail

The 1930's: Airmail's Heyday

By the 1930's, aviation routes were established worldwide. Even some intercontinental routes were in regular use. Collecting airmail covers had reached a fever pitch. This, of course, means there is a tremendous volume and variety of material available from the 1930's. As a result, this section is largely a picture-book of the type of material that is available. Some of this mail, for instance mail of many domestic airmail events, is very common and of limited interest, and therefore usually inexpensive. Other airmail, such as many Zeppelin flights and all catapault flights, is scarce and in high demand. So, such mail is *very* expensive. I try to show a little of both kinds of material. But, since the rarities often cannot be seen elsewhere, I placed greater emphasis on them.

The issues of rates and routes are not of major interest for the 1930's. Airmail rates were quite static, due largely to the depression of the period. As new intercontinental routes were opened, the new airmail rates are of some interest. Routes in Europe went everywhere. During the 1930's, over a half-dozen airlines incorporated Budapest into their European services. First flight mail of the era is interesting and generally inexpensive. Other than these first flights, the specific routes used to deliver airmail within Europe are essentially unidentifiable.

If there is a common thread to 1930's airmail, it has to do with the variety of conveyances used. We'll see mail carried on balloons, airships, airplanes of every variety, and even gliders. I've chosen as a starting point Hungarian mail carried on Zeppelin flights, since that service came to Hungary for the first time in 1930, at the very beginning of the decade we're reviewing.

Hungarian Mail for Zeppelin Flights

The origins of the Zeppelin-type airship are well-known and don't bear repeating here. The most famous of these dirigibles was the LZ-127, or *Graf Zeppelin*. This 105,000 m^3 giant was launched from the Zeppelin company's facility at Friedrichshafen, on Lake Constance, on September 18, 1928. Its first mail-carrying flight was to the United States and began on October 11. During its nine years of service, the Graf Zeppelin carried airmail and passengers primarily within Europe and to South America.

The Graf Zeppelin was owned by a private "for profit" company and only carried airmail on the basis of contracts that provided the required surcharge. The nations that made such arrangements are generally called "treaty

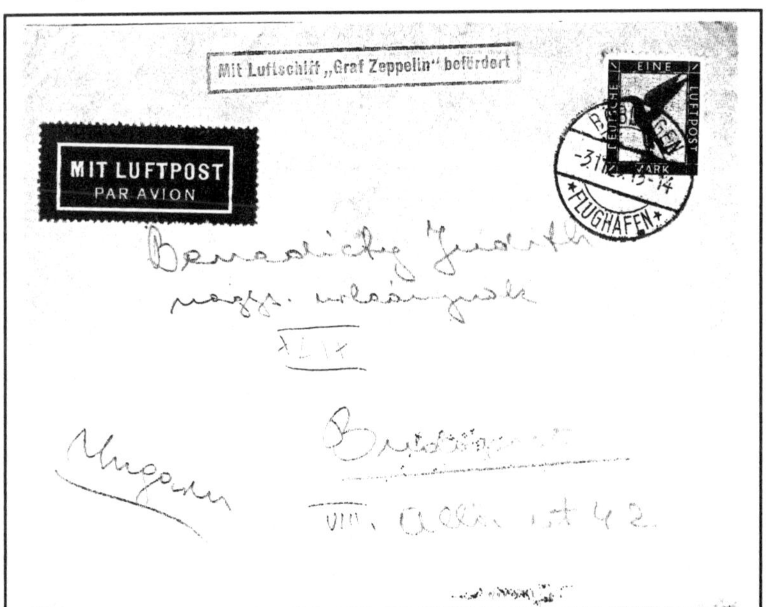

Figure 5-1 Zeppelin Mail of 1929 Addressed to Hungary

states." Hungary became a treaty state in 1930. Therefore, no mail with Hungarian postage was carried earlier. Figure 5-1 shows a German cover addressed to Hungary that was flown on a Böblingen-Friedrichshafen flight within Germany on November 3, 1929. The remainder of the trip to Budapest was made by regular airmail. Such mail is very scarce.

The mail we are talking about in this chapter is often called "connecting flight" mail. With the one exception of the *Hungary Flight* of 1931, which has a separate chapter devoted to it, the Graf Zeppelin did not come to Hungary. Therefore, mail intended for the Zeppelin was carried by connecting flights, usually to Berlin or Friedrichshafen, by regular airmail before being put aboard the airship.

For reasons that are uncertain, Hungarian usage of Graf Zeppelin airmail was much lower than that of many other treaty states. As a result, it is comparatively scarce and expensive. In this book, I have avoided discussing prices. I have talked about scarcity. In a "supply and demand" marketplace, scarcity and price are not necessarily correlated. Some very scarce 1922-27 airmails have been known to sell for very reasonable prices.

Hungary's Zeppelin mail has a much broader audience than just Hungarian airmail collectors. The demand is high for all Zeppelin mail. The combination of demand and scarcity means that all Hungarian Zeppelin connecting flight mail is expensive and some is *very* expensive.

Before looking at the variety of material available, a note on the role of the Sieger *Zeppelin Post Katalog* is required. That catalog is like a "bible" to most Zeppelin collectors. It lists all the flights and provides good coverage of German mail on those flights. But, the amount (or even the existence) of mail from a given treaty state for any specific flight is not generally available. As a result, Sieger's listings of *Vertragsstaaten* (treaty states) mail is dependent upon what has been reported to the cataloger.

A high point in any exhibitor's life is being able to say a certain piece of Zeppelin mail in their exhibit is "not listed by Sieger." With each new edition of the catalog, more such items become listed; but, additional new items continue to be found.

Based on the above comments, I have decided not to attempt yet another listing of the

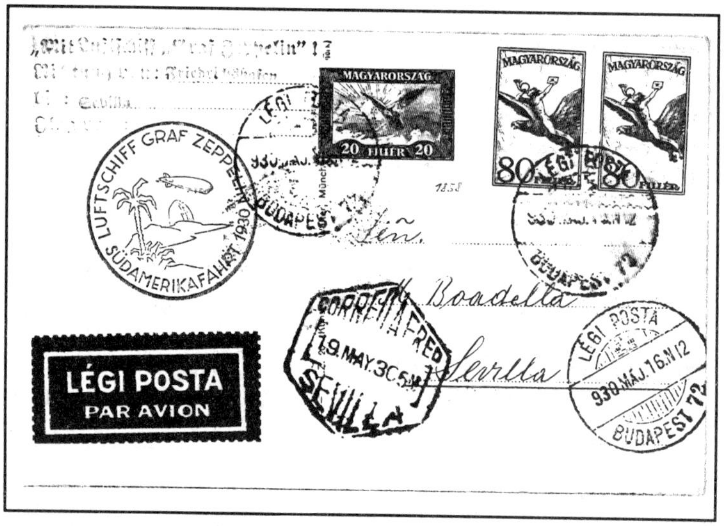

Figure 5-2 Hungarian 1930 Pan American Flight Mail

Figure 5-3 1930 Brazil Pan-American Flight Mail to Hungary

Hungarian mail carried by the Graf Zeppelin. Refer to Sieger, or your favorite other catalog, for the flight information you need.

1930 was the first year that Hungarian mail was accepted on the Graf Zeppelin. Usage started quite slowly, and most 1930 Hungarian Zeppelin mail is much scarcer than mail of the 1931–33 period.

The first flight I will discuss is the 1930 Pan America flight. This flight took the airship from Europe to South America, then to North America, and finally back to Europe. Hungarian mail carried on this flight is very rare. The only such mail I have seen is from the Friedrichshafen to Seville first leg of the flight. Figure 5–2 shows such a postcard.

Now, I must tell a personal story. In 1989, I exhibited my collection at a major Washington, DC exhibit. A friend, who specializes in the 1930 Pan American flight also exhibited there. Our exhibits happened to be across the aisle from each other, and we each showed virtually identical Hungarian cards flown to Seville -- the card in Figure 5–2. We both brought along a then current German auction catalog showing a very similar card for sale, with the assertion that "only two were flown"! The moral of the story is this: don't believe statements about the number of treaty states pieces flown on a given flight.

Figure 5–3 shows another aspect of how I choose to collect airmails. It is a postcard to Hungary from a member of the Hungarian legation in Rio de Janeiro. It was flown by the German-controlled Brazilian airline *Condor* from Rio to Recife where it was put aboard the Zeppelin. At Lakehurst, NJ the diamond–shaped *Europe Pan–America Round Flight* handstamp was applied. From the U.S. it was carried to Germany, via Spain. It then went to Budapest via normal airmail completing the "first direct South America–Europe airmail service" as the correspondent terms it. I have always enjoyed searching out airmail *to* Hungary. Graf Zeppelin mail from South America addressed to Hungary provides a particularly challenging search.

Figure 5–4 1930 Switzerland Flight Zeppelin Mail

The final 1930 Zeppelin flight I will show you, in Figure 5–4, is the October 12 *Switzerland Flight* to Basel. It falls into the category of "unlisted by Sieger" -- at least through the 20th edition of that catalog. I also show it to

Figure 5–5 The *Graf Zeppelin* Approaching the *Malygin*

Figure 5-7 Zeppelin Münster Flight -- Signed by Captain

make another point. Hungarian Zeppelin mail carried on flights to Switzerland is extremely expensive. Actually, *any* Zeppelin mail carried on flights to Switzerland is extremely expensive. Such mail is much sought after by the large contingent of "well-heeled" Swiss airmail collectors. So, as a result of demand more than scarcity, this class of Zeppelin item is very hard to add to any Hungarian airmail collection. I'll show a couple of more examples later in the chapter.

The first Graf Zeppelin flight of 1931 was the Hungary Flight. The next chapter, in its entirety, deals with that one flight. Here, I will continue showing you other Hungarian Graf Zeppelin mail that I hope will be of interest.

Two early season flights of the Graf Zeppelin are relatively common with Hungarian dispatches. One of these is the *Ostseejahr–Rundfahrt* ("Trip Around the Baltic Sea") of May 12–15. It is of special interest because collecting multiple destinations is possible. The second, and arguably the most common Hungarian connecting flight, is the Austrian flight of July 12, from Friedrichshafen to Vienna.

In July 1931, the Graf Zeppelin made a much-heralded trip to the Soviet Union and the Arctic. The highlight of the trip was lowering mail to the Soviet icebreaker *Malygin*. Figure 5–5 depicts the Zeppelin approaching the ship near Hooker Island, Franz Josef Land, about 80° N latitude.

Here again multiple destinations are possible. Hungarian mail flown Friedrichshafen–Berlin (arrival July 24) and Friedrichshafen–Leningrad (arrival July 25) is collected. The real prize though is mail lowered from the airship to the icebreaker Malygin on July 27. It is probably no scarcer than the land destinations, but demand is much greater. Figure 5–6 shows such a piece of mail.

Figure 5-6 Zeppelin Mail to the Soviet Icebreaker Malygin

Figure 5–7 shows a letter carried on the August 15–16 Münster flight within Germany. It illustrates two points. Hungarian mail of the lesser-known German flights is quite scarce and can be very interesting -- though there is less demand than for mail of the more spectacular flights. Also, this cover bears

Figure 5–8 Lausanne Flight of 1931

two signatures. While most of us aren't signature collectors, mail with the signature of a pilot adds both interest and information to a display of airmail. This cover is signed by the captain of the Graf Zeppelin, Heinrich Bauer, and also by Charles E. Rosendahl, a U.S. Navy officer and key official in the American airship program.

Figure 5–8 shows a card carried to Lausanne on the Switzerland flight of August 27. As with all flights to Switzerland, demand is particularly high. Note: at the left of this card is a clearly visible handstamp showing a Zeppelin and the words *Zeppelin Utján* ("On the route of the Zeppelin"). This was used on most, but not all, Hungarian Zeppelin mail from 1931–34.

Late in the 1931 season, the Graf Zeppelin began its series of passenger and airmail–carrying flights between Europe and South America. These continued until all Zeppelin travel ceased in 1937 following the Hindenburg disaster. The first and second South America flights of 1931 are the commonest of all the Hungarian Zeppelin material to South America. Even in these cases, some very elusive varieties are available.

Figure 5–9 shows a piece of August 31 drop mail to the Cape Verde Islands from the 1st South America flight. Note that the cachet shows the airship above the ocean. On the more common mail to Brazil, the cachet shows the Zeppelin docked on land. On its first flights, the Graf Zeppelin went only as far as Recife, Pernambuco state, in northeastern Brazil. A special Condor flight carried the Zeppelin mail to Rio de Janeiro, and returned with mail for dispatch to Europe.

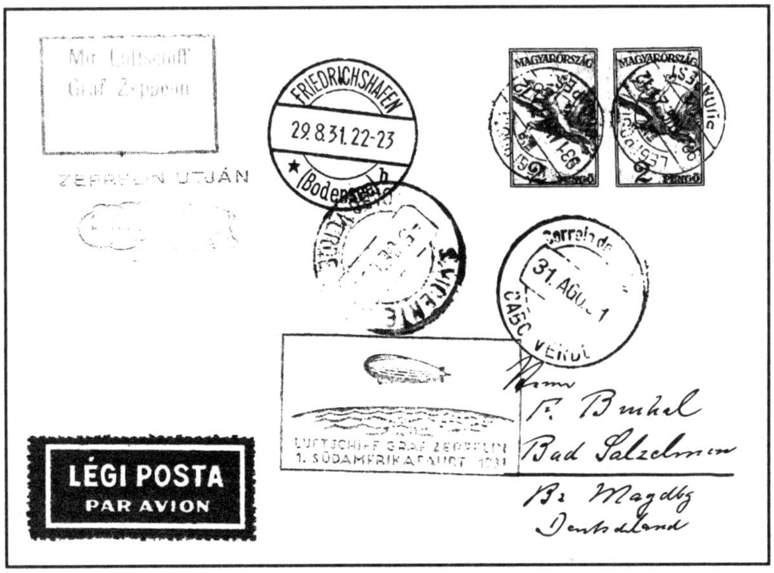

Figure 5–9 Drop Mail to Cabo Verde -- 1st SA Flight

Figures 5–10 to 5–12 show three different pieces of mail, all flown on the *2nd South America Flight* of 1931. Figure 5–10 is another card dropped to Cabo Verde enroute to South America. On the 2nd flight there was no special cachet for the drop mail. This card is particularly interesting because there was no previous record of Hungarian drop mail of the 2nd South America flight.

Figure 5-10 Drop Mail to Cabo Verde -- 2nd SA Flight

Figure 5-11 Second SA Flight of 1931 to Rio de Janeiro

Figure 5-11 is a card to Rio de Janeiro also carried on the 2nd South America flight. It is much more typical and common than the previous piece. Note the red bilingual handstamp at the lower left. It says that, "The Graf Zeppelin will wait four days for your reply in Pernambuco."

Figure 5-12 is Brazilian mail carried on the return leg of the 2nd South America flight. An interesting sidelight to a Hungarian airmail collection is South American Zeppelin mail destined for Hungary. Several South American countries issued special Zeppelin stamps, and used Zeppelin flight cachets. Typically in such mail, look for a Friedrichshafen receiving mark. That confirms on which flight the mail was carried. The mail was usually carried the remainder of the trip to Hungary by regular airmail.

The card in Figure 5-12 shows a Brazilian airmail stamp overprinted *2.500 Reis* specifically for the 2nd South America flight; a Rio postmark of September 23; two *Condor-Zeppelin* handstamps; and a Friedrichshafen receiving mark of September 28.

The 1932 Graf Zeppelin schedule was dominated by nine South American flights. Remember, these flights were numbered each season. Therefore, there is a *First South America Flight* for 1931, 1932, 1933, etc. I will show you only two examples of 1932's Hungarian Zeppelin mail.

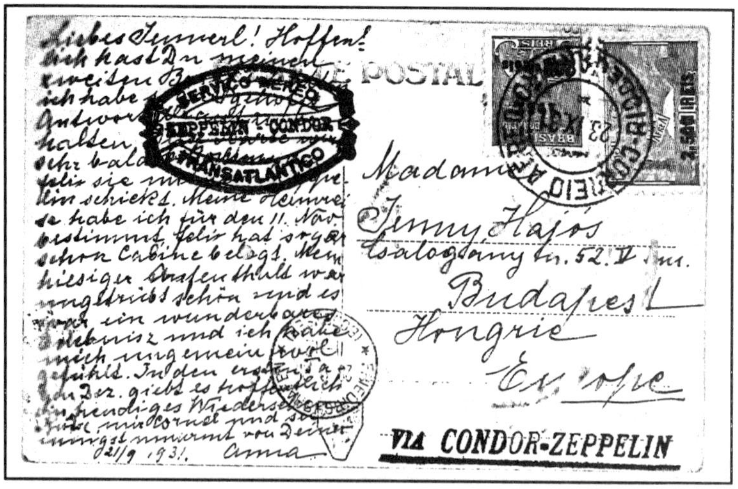

Figure 5-12 Brazil Mail to Hungary -- 2nd SA Flight

The most important European flight of the year was the July 31–August 2 flight to *LUPOSTA*, a major stamp exhibition in the free city of Danzig. Figure 5–13 shows a piece of Hungarian mail dropped by the Zeppelin enroute to Danzig at Rønne, on the Danish Baltic Sea island of Bornholm. While this drop mail is not listed by Sieger, it is noted by Keszy–Harmath is his specialized listing of Hungarian Zeppelin mail.

Figure 5–13 Drop Mail to Rønne in the Baltic Sea

The other 1932 Zeppelin flight mail is a commercial cover -- franked only by a postage meter. Figure 5–14 shows this cover, flown on the 7th 1932 South America flight of September 26 to October 4. Some Zeppelin collectors specialize only in commercial mail carried by the airship. Such mail is extremely scarce compared to the typical philatelic mail carried aboard the Zeppelins.

1933 had several interesting special flights, in addition to a series of regular South America flights. The Sieger catalog, through the 20th edition, does not attempt to list and price all the Hungarian airmail flown on the Graf Zeppelin's regular South America flights. Keszy–Harmath, though, does do that.

An interesting aspect of Hungary's South America Zeppelin mail of 1933, is that it often bears an additional cachet. In addition to the Hungarian "Zeppelin Utján" handstamp, and the handstamp for the specific flight, this mail often shows a cachet of a special *Anschlussflug* ("Connecting Flight") from Berlin to Friedrichshafen. These flights brought much of the treaty states mail, as well as some German mail, to the Zeppelin's departure point. This was important because Friedrichshafen was out of the mainstream of international airmail routes. In my opinion, this additional cachet gives the mail of 1933 a crowded and gaudy "look."

The first 1933 flight of the Graf Zeppelin was a test flight within Germany. Mail was dropped from the airship at the Dresden airport. Hungarian mail on this *Germany Flight* is very scarce. Figure 5–15 shows an example.

Figure 5–14 Commercial Mail Flown on 7th SA Flight of 1932

Figure 5-15 Germany Flight of 1933

Another flight within Europe that is very popular is the Rome flight of May 29–30. Figure 5-16 shows Hungarian mail carried on that flight. Mail of any of the treaty states, carried on the *Italian Flight*, is in great demand by Zeppelin collectors.

The final 1933 flight that I will discuss, is known as the *Chicago World's Fair Flight*. This was again a Pan-American flight that went from Europe to Brazil (Recife and Rio de Janeiro); and then to North America; finally returning to Germany via Seville, Spain.

In the U.S. the Zeppelin made stops at Miami, Akron, and then at Chicago during the world's fair we know as the *Century of Progress Exposition*. The dates of this flight were October 14 to November 2. Akron was included in the itinerary primarily because it was the home of the Goodyear Tire and Rubber Company. It was the principal manufacturer of airships in the U.S. So, in Akron there were hangar and maintenance facilities available.

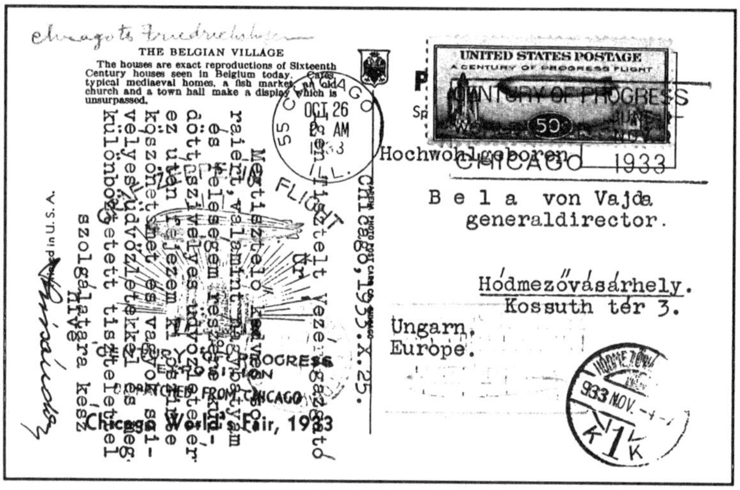

Figure 5-17 Chicago-Hungary Flight of 1933

The Zeppelin actually made only a quick trip from Akron to Chicago, returning to Akron the same day, October 25.

From a Hungarian collector's point-of-view, there are five Hungarian dispatches that are collectable (two to Brazil and three to the U.S.), and three from the U.S. to Hungary via Friedrichshafen. Before you set your sights on all eight, I should note several of these have never been recorded, and that even the most common -- Hungary-Chicago and Chicago-Hungary -- are extremely scarce. The

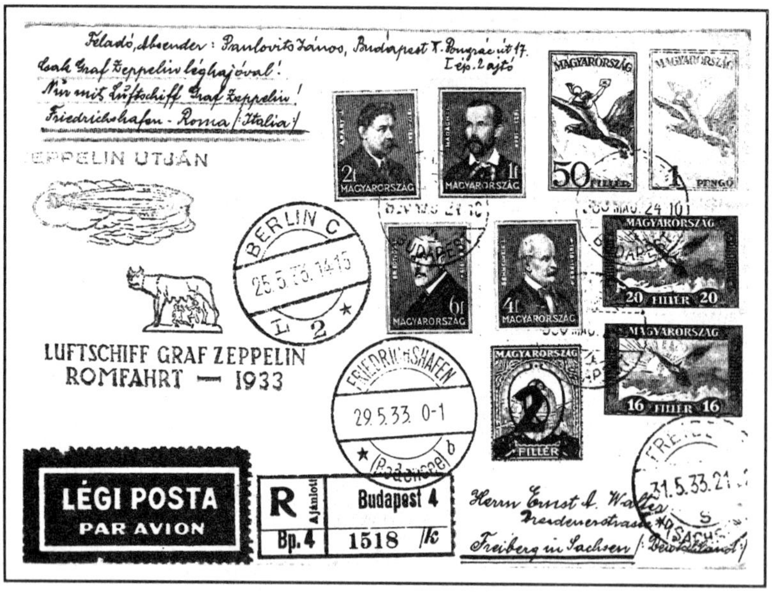

Figure 5-16 Graf Zeppelin Italy Flight to Rome

Figure 5-17 Hungarian Dispatch for the Chicago Flight

Chicago Flight mail is arguably the most expensive of any Hungarian Graf Zeppelin connecting flight mail.

Figure 5-16 shows a card flown from the U.S. to Hungary. Figure 5-17 show an example from Hungary to the U.S.

Beginning in 1934, many of the Graf Zeppelin flights did not have special cachets, and the quantity of philatelic mail decreased greatly. This meant that the quantity of collectable mail also decreased.

The most popular covers for many collectors are those which had cachets. Included among these is mail from the 3rd South America flight of the year, called the *Argentine Flight* because for the first time the Zeppelin flew all the way to Buenos Aires. This flight took place from June 23 to July 6. Again the Switzerland flight is of interest. This July 10 flight features mail dropped to the Zurich airport. Figure 5-18 shows an example.

Another flight cover with a special cachet is the 12th South America flight. Also known as the *Christmas Flight*, this last trip of the year took place from December 8-19.

Some collectors seek out the more ordinary, uncacheted Zeppelin mail of 1934-36. Most of this is commercial mail or diplomatic mail that often bears high frankings and has other interesting features. Figure 5-19 shows an example. This envelope was sent from the Brazilian legation in Budapest to their Foreign Minister in Rio de Janeiro. It was flown on the 6th South America Flight of August 18-28. The double-ring *Deutsche Luftpost / Europa-Sudamerika* handstamp on this mail has a code at the right indicating where the mail was put aboard the Zeppelin. In general, the *a* code indicates Berlin and the *b* and *c* code indicates Friedrichshafen. The code found in Figure 5-19 is a *c*.

Figure 5-20 shows an example flown on the *6th Graf Zeppelin South America Flight of 1936*. During 1936, the airship Hindenburg also made several flights to South America. As a result, there are three numbering systems in use for these flights -- the "nth" flight of the

Figure 5-18 Switzerland Flight of 1934

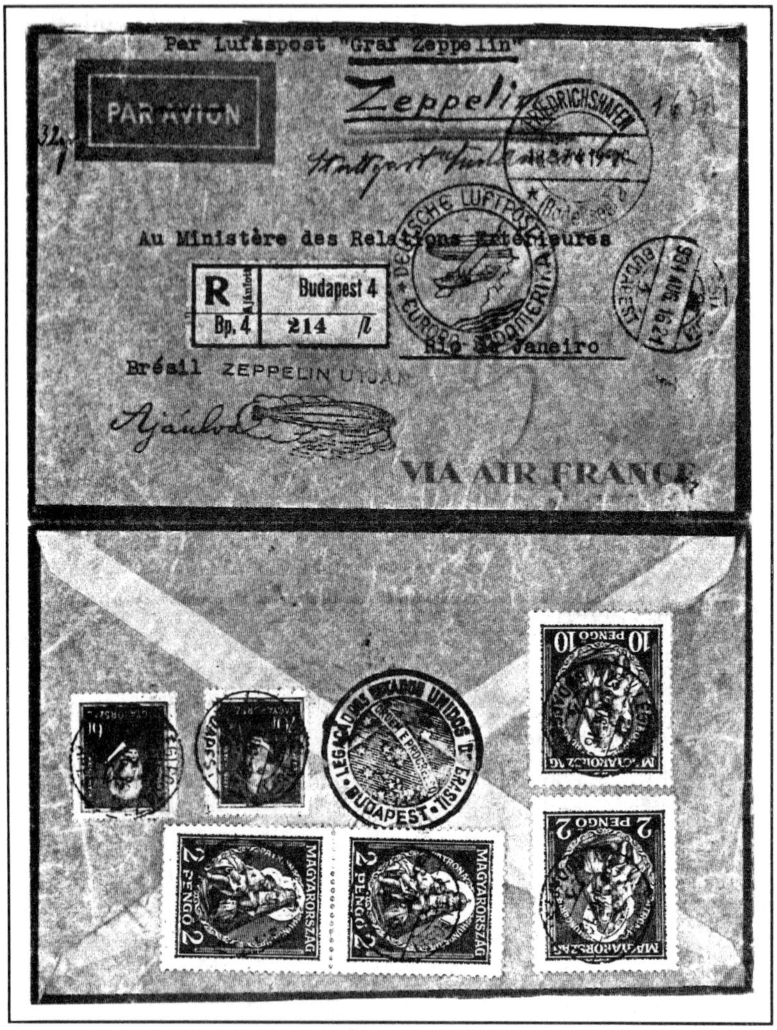

Figure 5–19 Sixth South America Flight of 1934

contract between Hungary and the Zeppelin company expired at the end of the 1936 flying season. Therefore, all Hungarian airmail to South America was flown by airplane beginning in 1937. Contradicting this, Keszy–Harmath lists Hungarian mail carried by the Hindenburg on the only 1937 flight to South America made before its crash in Lakehurst, NJ. Also, he lists Hungarian mail for the first (and only) two Graf Zeppelin South America flights of 1937. I have never seen any such mail.

Following the Hindenburg disaster, all commercial flights of the Zeppelins ceased. The Graf Zeppelin was formally decommissioned in 1940.

Hindenburg Mail. The LZ–129, known as the airship *Hindenburg*, was commissioned in March of 1936. During 1936, the Hindenburg made seven flights to South America and ten flights to North America.

Graf Zeppelin, of the Hindenburg, or overall. Therefore, this flight may be referred to as the *8th South America Flight of 1936* (overall) or the *6th Graf Zeppelin South America Flight of 1936*.

The *a* on the double–ring handstamp indicates this cover was flown by the Zeppelin from Berlin. It was carried from Rio de Janeiro to Buenos Aires by normal airmail.

I have been told by Zeppelin experts that the

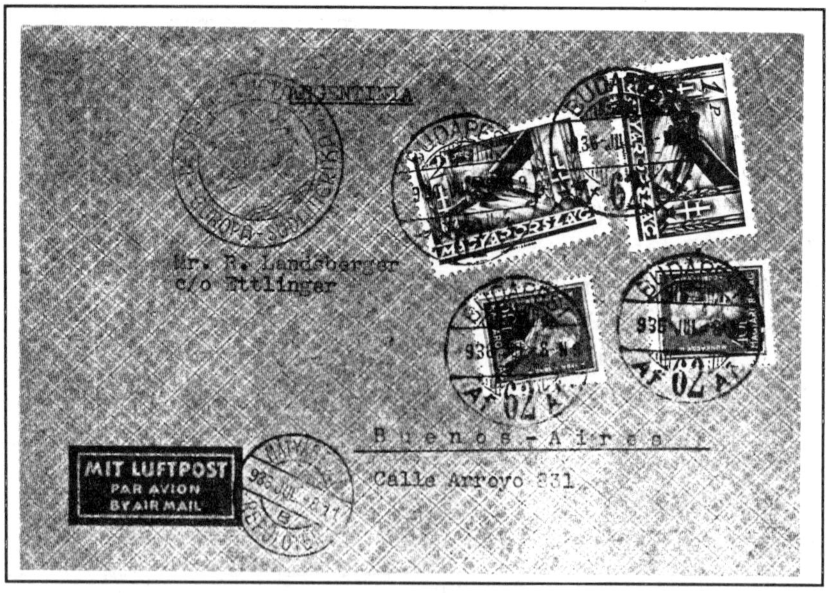

Figure 5–20 Graf Zeppelin's 6th SA Flight of 1936

Figure 5-21 The LZ-129 Hindenburg In Flight

Figure 5-22 First North America Flight of the Hindenburg

According to Keszy-Harmath, Hungarian mail was carried on each of these. With the exception of the *1st North America Flight*, mail from these flights is quite scarce.

After making one flight to South America early in 1937, the Hindenburg burst into flames while preparing to dock at Lakehurst, NJ on its first trip of the year to North America. After seeing the film of the Hindenburg disaster many times, what amazes me most is that *anyone* survived. No Hungarian mail was found among the mail salvaged from the wreckage.

Figure 5-21 shows a photo of the Hindenburg in flight. Figure 5-22 is an example of Hungarian mail carried on the *1st North America Flight* of 1936.

The only European flight of the Hindenburg to carry Hungarian mail was the *Olympic Flight*. The LZ-129 flew over the Olympic Stadium in Berlin on August 1. Figure 5-23 shows Hungarian mail carried on the Olympic Flight. This is the rarest and most sought after of Hungarian *Hindenburg* mail.

Figure 5-23 Hindenburg 1936 Olympic Flight

The Graf Zeppelin Hungary Flight

The Graf Zeppelin first flew over Hungarian territory during the *Balkan Flight* of 1929, but did not land. The first and only flight that actually landed in Hungary took place in late March 1931.

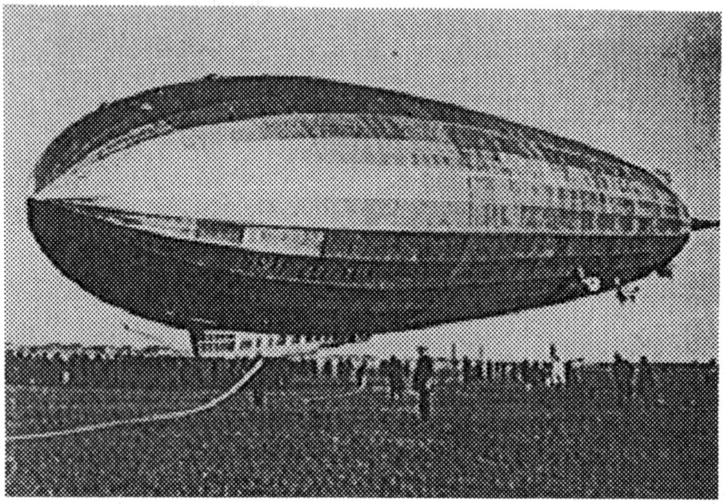

Figure 5–24 Graf Zeppelin Docked Near Budapest on March 29, 1931

The activities associated with the presence of the airship in Hungary were organized by the *Royal Hungarian Automobile Club*. Hungary issued two special stamps for use on the various phases of the flight. We know these stamps as Scott #C24–25. They were only valid for the Graf Zeppelin *Hungary Flight*. These stamps are not separately illustrated here, but can be seen on airmail in Figure 5–27 and several other illustrated covers in this chapter. Figure 5–24 shows the Graf Zeppelin docked near Budapest on March 29, 1931. The location of the Budapest dockings was the Manfred Weiss Works on Csepel island in the Danube.

The Graf Zeppelin *Hungary Flight* was completed in three phases. At 11PM on March 28, the Zeppelin left its base in Friedrichshafen for its *Flight to Hungary*. It carried mail dispatched from Germany and eight treaty states, including Hungary. Mail also was posted on–board the airship. It arrived at Budapest the morning of March 29.

The second phase of the flight was a planned trip around Hungary during the day on the 29th. The purpose of this *Round Flight* was to "parade the colors," i.e. let the Zeppelin be seen in all the major cities of Hungary. Mail was to be dropped to several of these cities. As a result of high winds, no dockings could be made and all but one mail drop was canceled. Mail was dropped only to Debrecen. The remainder of the mail was returned to Budapest. The mail carried on this *Round Flight* is the most common of Hungary's Zeppelin mail.

The Graf Zeppelin left Budapest late in the afternoon on the 29th. It carried Hungarian mail for the *Return Flight* to Germany. Mail was also posted on–board the *Return Flight*. The flight ended at Friedrichshafen on the morning of March 30.

Figure 5–25 German Mail Flown to Hungary

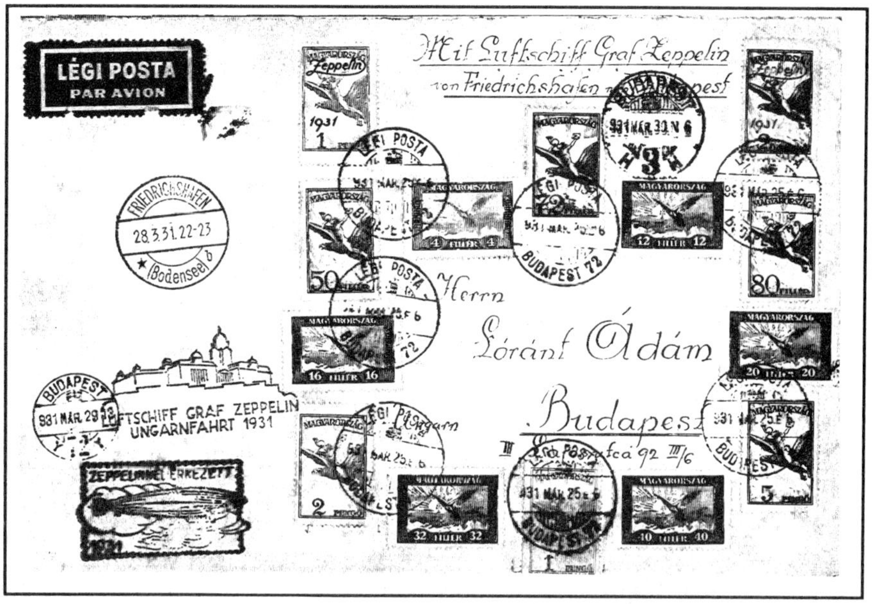

Figure 5–27 Hungarian Mail Flown on Flight to Hungary

The Stamps. The Graf Zeppelin stamps were issued specifically for the *Hungary Flight*. Their period of validity was only March 23–31, 1931. These stamps *are* the 1 Pengő and 2 Pengő airmail stamps of 1930, printed in *different* colors, with *Zeppelin / 1931* overprinted on them. About 60,000 of each value were printed.

The 1 Pengő value is printed in orange with a black overprint. The 2 Pengő value is printed in violet with a green overprint. Imperforates exist. There also exist examples, in the colors of the Zeppelin stamps, without the *Zeppelin / 1931* overprint. These are not ,recognized by most catalogs, and probably represent materials stolen from postal archives at the end of World War II.

These stamps were only valid to pay the Zeppelin mail surcharge, but they were *not* required for that purpose. Many collectors seek out Zeppelin *Hungary Flight* mail with Hungarian postage that does not include either of the Zeppelin stamps.

For the *Flight to Hungary* and the *Return Flight* to Germany, the Zeppelin surcharge was 2P for postcards, 3P for letters. For the *Round Flight*, the Zeppelin surcharge was 1P for postcards and 2P for letters. The regular airmail postage rates also applied.

Flight to Hungary. On its *Flight to Hungary*, the Graf Zeppelin left Friedrichshafen at 11PM on March 28 and arrived at Budapest at 8:30AM on the 29th. The landing site was the airfield of the Manfred Weiss Airplane Factory on Csepel, the large island in the Danube just south of Budapest.

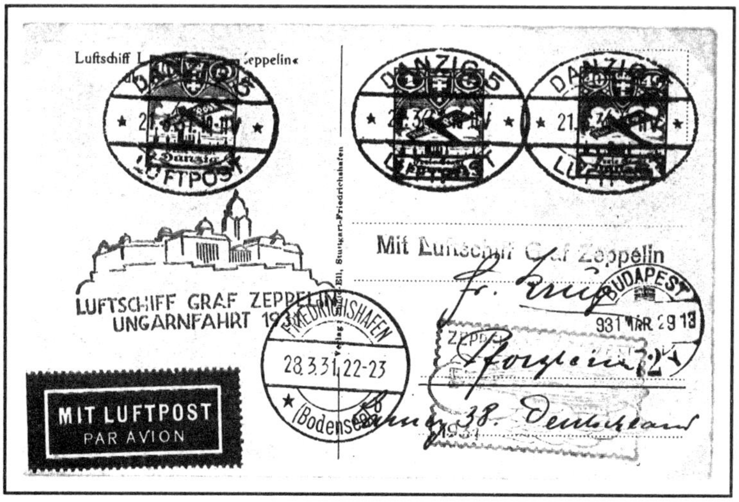

Figure 5–26 Danzig Mail of the Flight to Hungary

The Zeppelin carried 3,406 postcards and 1,574 letters to Budapest. The bulk of this was German mail either postmarked at Friedrichshafen or on–board. Figure 5–25 shows an example of the former. Figure 5–28 shows the Zeppelin's on–board cancellation. This mail all bears a red cachet picturing the Royal Palace at Buda with the words *Luftschiff Graf Zeppelin / Ungarnfahrt 1931* below.

On receipt, a red receiving mark with a picture of the Zeppelin and the words *Zeppelinnel Érkezett / 1931* was applied, as was a standard

Budapest 72 circular date stamp of March 29 with a time indication of *13*. These markings are seen on Figures 5–25 through 5–27.

The Zeppelin also carried mail of the treaty states. Mail of the following countries was carried: Austria, Danzig, Hungary, Liechtenstein, Luxembourg, Netherlands, Saar, and Switzerland. Figure 5–26 shows an example of Danzig mail carried on the *Flight to Hungary*.

Luxembourg mail has both the stamps of Luxembourg and Germany. These mixed frankings are normal for Luxembourg Zeppelin mail. This is by far the rarest of the treaty states mail carried on the *Flight to Hungary*. I have never seen an example.

Many in Hungary were not aware that Hungarian mail could be forwarded by connecting flight for the Zeppelin's *Flight to Hungary*. Therefore, Hungarian mail carried on this flight is less common than that of some of the other treaty states. It is on this Hungarian mail that the earliest postmarks (March 23–25) are found on the Zeppelin stamps. Figure 5–27 shows a large, philatelicly overfranked cover carried by the *Flight to Hungary* that bears the Zeppelin stamps plus the entire 1927–30 airmail issue.

A small number of cards with mixed Hungarian and German frankings are known from the *Flight to Hungary*. All are similar, and were probably created by a passenger intending to have the Hungarian stamps canceled on-board. Since this was not legal, German stamps were added and canceled on-board. The German postage paid the entire fee. As a result of special handling, the Hungarian Zeppelin stamp was tied by the receiving mark on arrival at Budapest. Figure 5–28 shows an example of this very scarce variety.

At least one other apparent mixed franking exists. This, unlike the Germany–Luxembourg and Germany–Hungary mixed frankings discussed above, actually represents a *remailing*. The recorded example is a piece of Italian mail that was carried to its addressee in Italy. The envelope was later remailed aboard the Graf Zeppelin, where German stamps were applied and canceled. The Italian postage has no relationship whatsoever to the Zeppelin flight. I would consider such a cover just a "curiosity" item.

Round Flight. The *Round Flight* took place during the day on Sunday, March 29. It was intended to give the people in all the major cities of Hungary an opportunity to see the Graf Zeppelin. The plan was to drop mail at several of the larger cities, including: Debrecen, Szeged, Kaposvár, Pécs, Szombathely, and Székesfehérvár. Due to high winds, this was not possible. The only mail drop was made at Debrecen. Here, the mail intended for the other cities was also dropped. Much of the mail for the other cities was prepared by the Royal Hungarian Automobile Club.

In addition to mail, the *Round Flight* carried 30 passengers who paid 375 Pengó each for the privilege. Six or seven postal officials also made the trip.

The *Round Flight* carried 15,359 postcards and 17,070 letters. The number carried for the entire flight, versus those to be dropped enroute, is unknown. The large

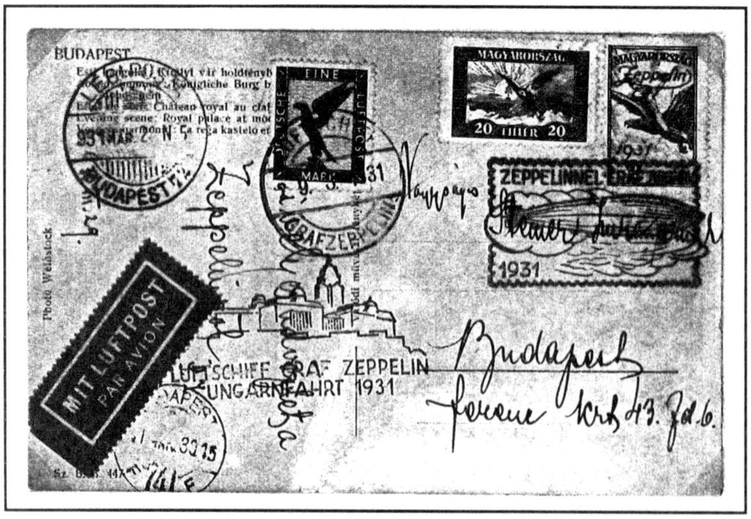

Figure 5–28 German–Hungarian Mixed Franking

quantity of mail carried means that it can still be found at reasonable prices. The *Round Flight* mail is by far the least expensive of Hungarian Graf Zeppelin mail. But, there are several specialty items from the *Round Flight* that are very scarce and in demand. These bring rather substantial prices.

Figure 5–29 Round Flight Mail -- Cachet Type 1

All the mail of the *Round Flight* was handstamped with a blue circular cachet. Apparently two of these handstamps were used producing slightly different forms of the cachet. For most examples, differentiating between the two types of the *Round Flight* cachet is simple.

Figure 5–30 Round Flight Drop Mail -- Cachet Type 2

Let's number the ribs shown on the airship from top to bottom and call them rib 1 to rib 6. The primary rule is this: if rib 5 (the second rib up from the bottom) only shows near the front of the airship, it is type 1; if rib 5 consists of dashed lines going most of the length of the airship, it is type 2. This distinction is readily apparent in Figure 5–29 that has a type 1 cachet, and in Figure 5–30 that has a type 2 cachet.

This rule should suffice for identification of 95% of the cachets. The only exception is where the lower half of the cachet is very weakly applied. In that case, a secondary rule may be of some help. This rule is: if rib 2 (the second from the top) consists of only two short segments in the middle of the airship, it is a type 1 cachet; if the line segments that make up rib 2 go to the ends of the airship, it is type 2. These distinctions can also be seen in Figures 5–29 and 5–30. To date, I've found only one example that was so poorly applied that neither of these rules permitted differentiation.

As previously noted, the Hungarian mail of the *Round Flight* can be divided into two categories. The first is mail carried on the entire flight. It bears the March 29 receiving mark of *Budapest 741* with a time–stamp of *20*. Much of this mail consists of postcards and envelopes prepared by, and addressed to, the Royal Hungarian Auto Club. Figure 5–29 is an example.

The second category is mail dropped to Debrecen. This mail bears a March 29 *Debrecen 2* receiving mark with the time–

stamp of *13* or *14*. Figure 5–30 is an example.

Figure 5–32 The Rarest of the Hungary Flight Varieties

The quantities of mail returned to Budapest, versus that dropped to Debrecen is unknown. The Debrecen drop mail seems to be somewhat scarcer in the West, while Keszy–Harmath's pricing indicates it is somewhat more common in Hungary. In any case, the difference is not great.

A third, albeit rare, variety exists. This is on–board mail with German franking. It was carried on the *Flight to Hungary* with all its typical markings. The mail was returned to the Zeppelin and flown on the *Round Flight* but, it does not bear the blue Hungarian *Round Flight* cachet. It was postmarked again when the Zeppelin returned to Budapest by a *Budapest 741* with a time–stamp of *18*. Some "special handling" was obviously involved with this mail.

A few comments about quantities and distribution of the two types of blue, circular *Round Flight* cachet are necessary. The overall quantities of the two types of this cachet are roughly equal. For mail returned to Budapest, most of the postcards have type 1 cachets; most of the envelopes type 2. Exceptions (type 2 postcards or type 1 envelopes) are rather unusual.

For mail dropped to Debrecen, most of the mail was handstamped with the type 1 cachet. The type 2 cachet to Debrecen, on either postcards or letters, is quite unusual. Figure 5–30 shows an example of this scarce type, a drop card to Debrecen with the type 2 cachet.

The Keszy–Harmath catalog lists the above varieties. His indications of value show little difference in price, though clearly there are big differences in scarcity. This simply reflects the fact that most collectors don't concern themselves with these minor distinctions.

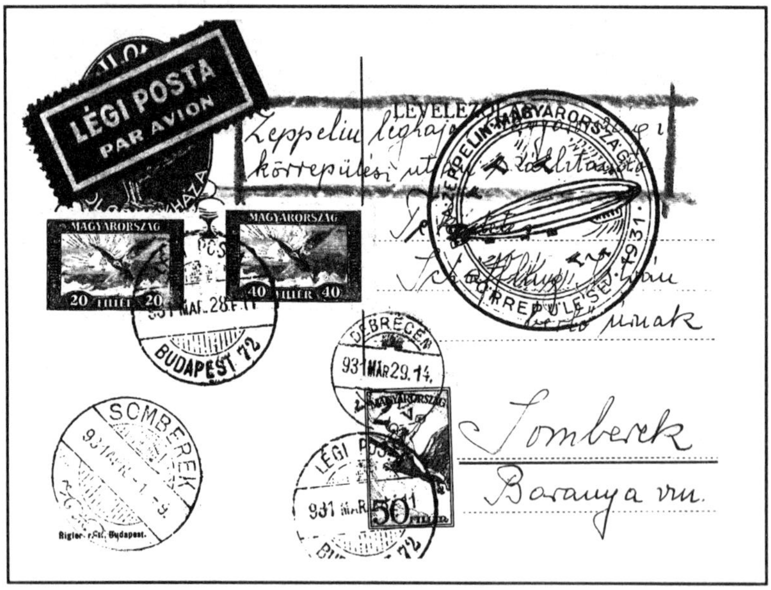

Figure 5–31 Round Flight Mail without Zeppelin Stamps

Another approach taken by Keszy–Harmath is listing and pricing *Hungary Flight* mail based on the stamps used. He distinguishes four varieties: mail franked with the 1P Zeppelin stamp; mail franked with the 2P Zeppelin stamp; mail franked with both Zeppelin stamps; and mail franked with no Zeppelin stamps. The latter is much scarcer and higher priced than mail with the Zeppelin stamps.

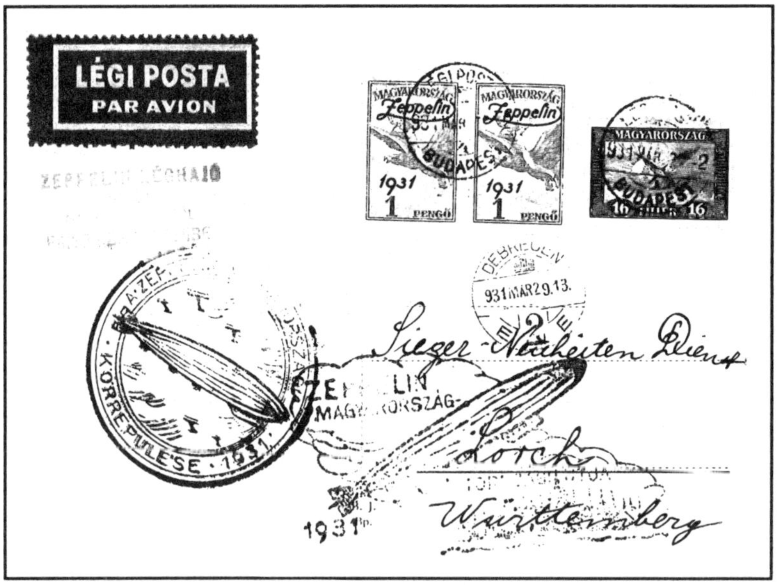

Figure 5–33 Mishandled Mail Dropped to Debrecen

Although the mail without Zeppelin stamps is gaining a following, the price difference is still much smaller than its relative scarcity. This again is due to the limited number of collectors who focus on Zeppelin mail at that level of detail.

Figure 5–31 shows an example of a postcard carried to Debrecen on the *Round Flight* without any Zeppelin stamps. Figure 5–32 shows a card carried on both the *Round Flight* and the *Return Flight* to Germany that also is not franked with Zeppelin stamps. This latter card combines the two rarest attributes of *Hungary Flight* mail, and is unlisted even by Keszy–Harmath.

The final example of *Round Flight* mail that I will show is a very unusual piece of mail that was erroneously handled. The card in Figure 5–33 was marked and franked for the *Return Flight* to Germany. The green cachet for that flight was even applied. It was inadvertently cacheted for the *Round Flight* as well; and was flown on that flight. Since it was dropped to Debrecen, it never made it to the *Return Flight* to Germany.

Return Flight. The Zeppelin left Budapest late in the afternoon of March 29. It arrived back at Friedrichshafen very early on the morning of the 30th. On the *Return Flight* it carried 2,722 postcards and 3,653 letters. It also carried 20 passengers who boarded in Hungary after paying 500 Pengó per ticket.

The green cachet for the *Return Flight* pictured the Zeppelin, flying in clouds, above the Hungarian Parliament building in Pest. This *Return Flight* cachet is also found in two types, again probably representing two handstamps used by different postal employees.

Figure 5–34 Return Flight Postcard with 6–Window Cachet

The two types are distinguishable by the number of windows in the gondola beneath the Zeppelin. [In the handstamp, the "windows" are represented merely by dots of color.] Type 1 shows six windows; type 2 shows five windows. Both types seem to be equally common, and there are no distinctions as to type of mail, as is the case with the *Round Flight* cachet.

Figure 5–35 Return Flight Mail with 5–Window Cachet

Figure 5–34 shows an example of the 6–window cachet; Figure 5–35 shows the 5–window type.

A receiving mark was applied to *Return Flight* mail at Friedrichshafen on March 30. The time–stamp is *6–7*.

Keszy–Harmath again lists the *Return Flight* mail by type of cachet and type of franking. As with the *Round Flight*, the scarcest frankings of all are those without any Zeppelin stamps.

Mail posted on board the Zeppelin during the *Return Flight* exists. It is not listed by Sieger, but is listed by Keszy–Harmath. The postage, of course, is German; and, the on–board cancellation was used. Figure 5–36 shows an example. This mail is very scarce, but due to its German franking is not of interest to many Hungarian collectors.

The final item that I will show from the *Return Flight* is a card that should never have been flown. Figure 5–37 shows a postcard franked for the *Round Flight*, with a private handstamp confirming this intention.

Apparently, it was posted too late to make the *Round Flight*. So, the green cachet of the *Return Flight* was applied and it was put aboard the Zeppelin with the mail for Germany. It was flown, and bears the Friedrichs–hafen receiving mark. It is the only underfranked Zeppelin mail of the *Hungary Flight* that I have seen.

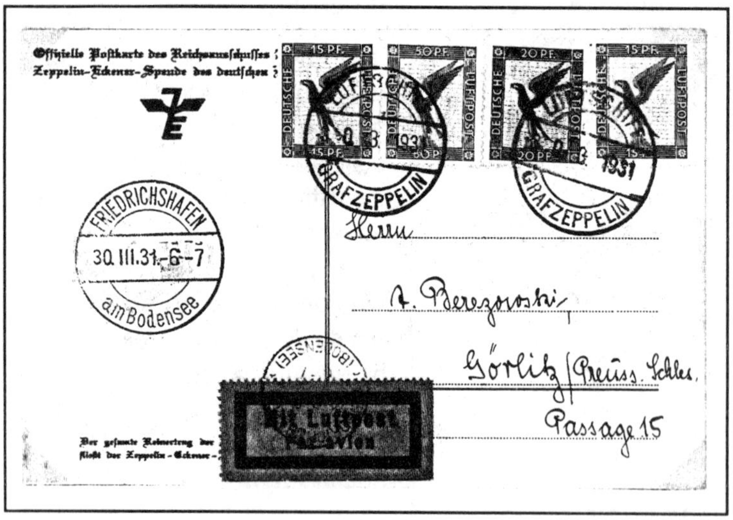

Figure 5–36 On–Board Mail of Return Flight

The Graf Zeppelin *Hungary Flight* made many Hungarians much more aware of the Zeppelin and its philatelic exploits. As a result, the amount of Hungarian mail on subsequent flights of the airship increased considerably from the 1930 levels.

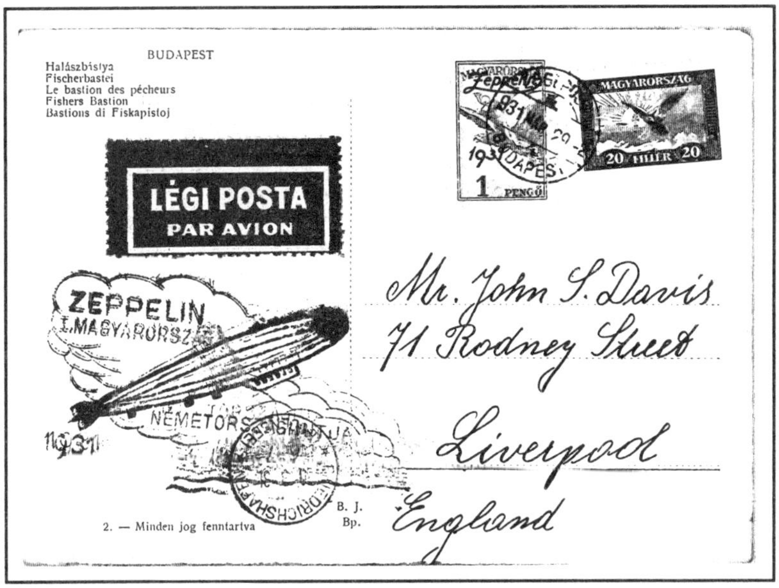

Figure 5–37 Underfranked Card of the Return Flight

Justice for Hungary and Star Flights

The Treaty of Trianon of 1920 distributed the greater part of Hungary's land and people to Romania, Czechoslovakia, and Yugoslavia. This was a major blow to Hungarian industry and transportation; but, the greatest bitterness was caused by the large number of ethnic Hungarians who found themselves separated from their homeland. Section 1 discussed the impact of the treaty on Hungary.

As a means of seeking publicity for the plight of Hungary, Lord Rothermere, the publisher of London's *Daily Mail*, offered a prize of £2,000 in 1931 (then US$10,000) to the first fliers to go nonstop from America to Hungary. On this side of the ocean, the idea of a trans–oceanic flight to Hungary was born in a neighborhood tavern in Toronto. To realize the goal, the meager contributions of the local people, which started the effort, had to be supplemented by a major fund–raising campaign. This financing came from Emil Szalay, a wealthy Hungarian–American meat packer in Flint, Michigan. George Endresz and Alexander Wilczek (also known as Alex Magyar) took up the challenge. Both were World War I Austro–Hungarian army pilots who had emigrated to America. As we shall see, Endresz died in a 1932 crash; but, Magyar lived a long life, passing away only in 1990.

Figure 5–38 shows two picture postcards. The first shows Endresz and Magyar (Wilczek); the second pictures their airplane where it landed in Hungary. It is signed by both George Endresz, and the flight's benefactor, Emil Szalay.

In June, Endresz and Magyar flew their Lockheed *Sirius* monoplane, named *Justice for Hungary*, from Detroit to New York. Here they campaigned, for the cause described by the name of their airplane, in the Hungarian communities of New York, New Jersey, and Connecticut. Then they flew on to Harbour Grace, Newfoundland, near the easternmost point in America. At 1:55 PM on

Figure 5–38 *Justice for Hungary* Pilots and Lockheed *Sirius* Monoplane.

Figure 5-39 Justice for Hungary Postcard Returned in Bulk to U.S.

July 15, they flew out over the Atlantic. After 25 hours and 40 minutes, they ran out of fuel and landed in a field in Bicske, fifteen miles short of their Budapest destination. Their total distance on this 12th non-stop crossing of the north Atlantic was 3,200 miles, farther than Lindbergh flew. Despite the forced landing, they were well within their goal of flying from America to Hungary.

I'll let Magyar describe their arrival in Hungary in his own words, from his 1940 book **We Dared to Dream**. *Somewhere over Györ the engine started sputtering ... Endresz was frantically switching fuel tanks ... it was too late. The propeller had stopped and we had to land. We were approaching a field much too fast when we noticed that a hay-wagon had turned into our path. Endresz turned the rudder hard left, but the landing wheel had already touched the ground. The plane bounced hard, then came to a stop at the edge of a cornfield ... I jumped out first, followed by Endresz ... Endresz ... said "Well, at least we did it!" Meanwhile, the driver of the hay-wagon, still unaware of his close brush with death, had come over to us. "Where are you coming from?" he asked in an earthy dialect. "From America," we replied. He looked at us puzzled. We could tell that his feelings were hurt. He thought that we were making fun of him.*

The great bulk of the mail carried by the *Justice for Hungary* flight consisted of special postcards that were imprinted with a picture of a dove and the inscription:

Detroit-Budapest
1 Dollar 1

in the normal stamp position. These were sold to raise money for the flight. Although 30,000 of these cards were printed, only 4,600 of them were sold and flown. Most bear the then current one cent green Benjamin Franklin stamp, which paid the domestic postcard rate. They are usually canceled in Flint, Michigan on May 14. The domestic postcard rate was used on the premise that the *Justice for Hungary Flight* itself was not an official airmail flight. Some of these postcards are also known bearing a 2¢ Canadian stamp, or a U.S. 5¢ stamp. The latter was the domestic airmail rate. A few of these cards, that were not franked with stamps and did not enter the U.S. postal system, were also carried on the flight .

All the *Justice for Hungary* airmail was postmarked on receipt by the three special green handstamps seen on several of the covers illustrated below. The receiving marks were: a standard *Budapest* 72 circular date stamp of

Figure 5-40 Delivered in Hungary without Postage Due

July 16; a circular marking picturing the airplane with English and Hungarian text; and three lines of text with basic information on the flight. Note that the mail was held three days while the special backdated markings were prepared. The use of these special receiving marks, in my mind, makes the receipt of the mail official in Hungary; thereby affecting its status regarding postage due.

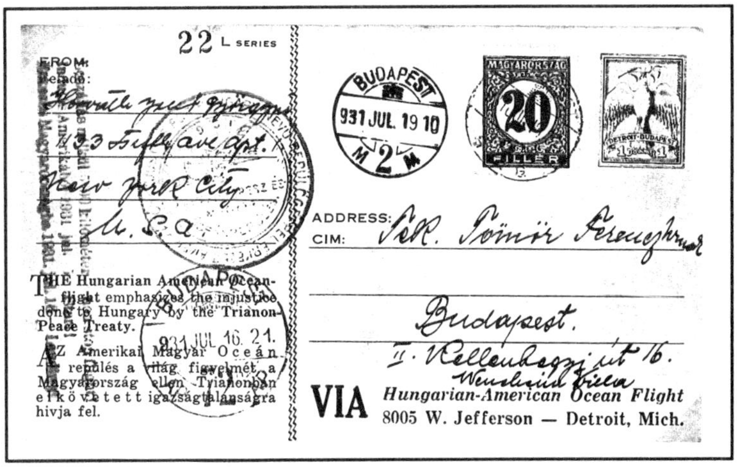

Figure 5-42 Unstamped Card Delivered in Hungary

Many of the postcards were returned to the U.S. in bulk, under separate cover. Postage due was never assessed on these. Figure 5-39 shows such a card.

Figure 5-43 Justice for Hungary Cover with 15¢ Franking

Some cards were delivered to addressees in Hungary. These were mostly franked with a 1¢ U.S. stamp, and so when they entered the Hungarian postal system were usually deficient in postage. Postage due of 32 fillér should have been assessed on these post-cards. Actually, some of this mail was delivered without postage due. Figure 5-40 shows an example delivered to Debrecen. Other examples are found with Hungarian postage due assessed. Figure 5-41 shows such a card delivered to Szeged with postage due.

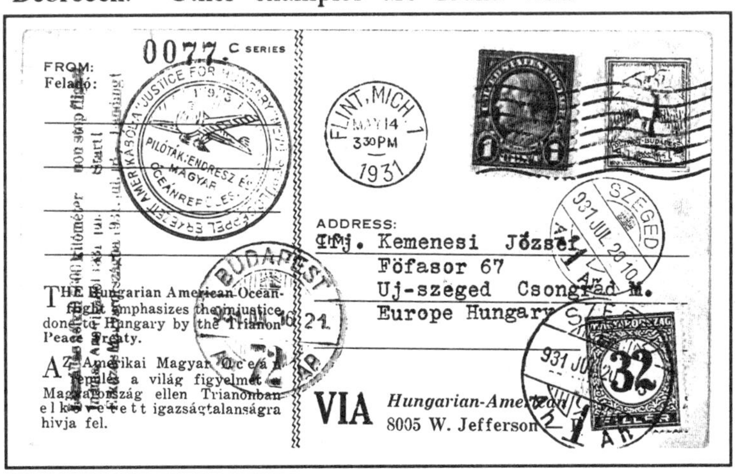

Figure 5-41 Delivered in Hungary with Postage Due

An interesting anomaly occurred with the handful of postcards that were carried on the flight without ever having been in the U.S. mails. The postage due assessment is double the deficiency (based on the then concurrent conversion rate for French francs). For the normal cards with one cent postage this amounted to 32 fillér -- to make up the equivalent of the international surface rate. But, the cards that were never in the U.S. mail entered the Hungarian postal system as unfranked domestic mail. Therefore, the postage due was only 20 fillér, despite having no stamps affixed! Figure 5-42 shows an example of this very unusual variety.

Mail carried aboard the flight, other than the special postcards, is very rare. This mail is usually franked with the correct 15¢ international airmail rate. Figure 5-43 illustrates an envelope of this type carried aboard the flight. It is signed by Alex Magyar.

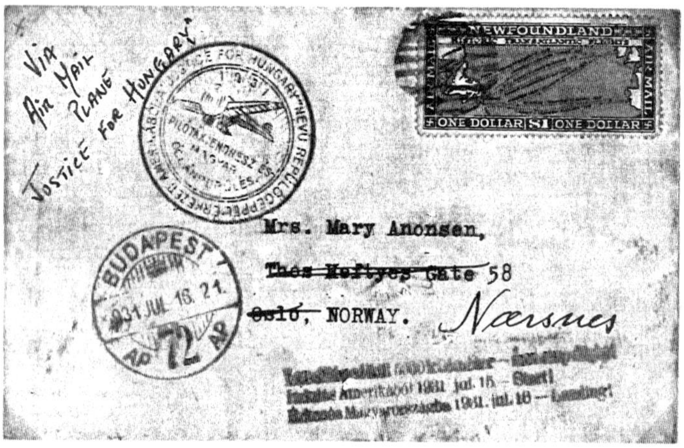

Figure 5-44 Justice for Hungary Newfoundland Mail

The final type of airmail carried by the *Justice for Hungary* was put on board in Newfoundland. These cards and covers, which only numbered about 20, are franked with the then current $1 Newfoundland airmail stamp. This mail was put aboard by W.H. Kennedy, the Harbour Grace post-master, without the approval of the Newfoundland Postal Minister. [Note: at this time Newfoundland was a separate British Dominion, not a part of Canada.] Figure 5-44 shows an example of New-foundland airmail carried on the flight to Hungary. This letter, after arrival in Hungary, was forwarded to its destination in Norway. It is backstamped in Oslo.

In the autumn of 1931, Endresz and Magyar flew to England to collect their prize from Lord Rothermere. In November, Magyar returned to the U.S. On the 13th, he was welcomed back to New York with a major reception hosted by Mayor Jimmy Walker. Endresz continued barnstorming in Europe with the *Justice for Hungary*. On May 21, 1932 he left Budapest for Rome to attend a conference of trans-Atlantic pilots. He crashed on landing in Rome and died along with his radio operator, László Bittay.

The Italian dictator, Benito Mussolini, provided a Fiat BR-3 monoplane that he named the *Giustizia per L'Ungheria* to continue the propaganda for revision of the Treaty of Trianon. This plane was flown to Hungary on June 15, 1932 by the Italian aviators Liberati and Constanzi. Mail with a special "Róma-Budapest" vignette was postmarked at Mátyásföld that day.

Star Flights. In September, the new "Justice for Hungary" airplane, piloted by a Hungarian named Lázsló Kálman, undertook the "Star of Savoy Air Tour." [The "Star of Savoy" was the emblem of the Italian royal family.] This 10-city tour is philatelicly divided in five flights that are usually called the **Star Flights**. In

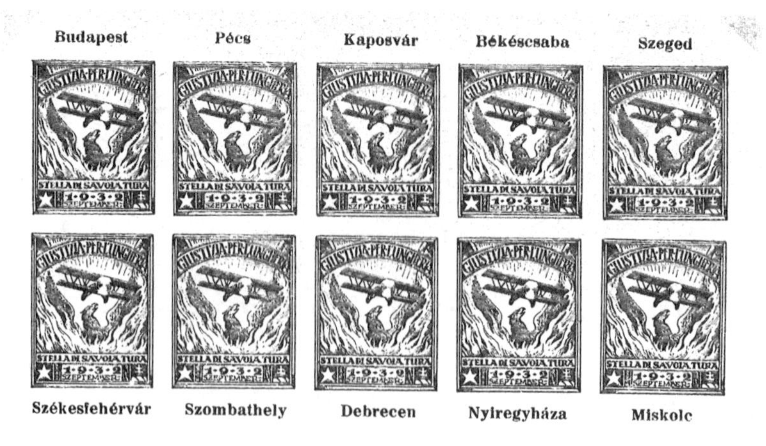

Figure 5-45 Star Flights Vignettes

Figure 5-46 Card Flown on Star Flight 1

addition to the "Star" reference in the official name of the air tour, the term *Star Flights* is especially appro–priate because the flights radi–ated out from Budapest in the form of a star. As we will see, a sixth flight was added in October. A sheet of ten green vignettes was printed with the names of the cities in the margin. Figure 5–45 shows an unused sheet of these vignettes.

Figure 5-47 Cover Flown on Star Flight 2 from Szeged

A handstamped cachet was prepared for each flight, with a code indicating the origin of each leg of the flight. A basic way of collecting this material is to find each of the five cachets with each of the origin codes. The codes are: Budapest = Bp; Mátyásföld = Mf; Pécs = P; Kaposvár = K; Szeged = Sze; Békéscsaba = B; Szombathely = Szb; Székesfehérvár = Szf; Debrecen = D; Nyiregyháza = Ny; and Miskolc = M. Note that legs beginning in Budapest could have either the *Bp* or *Mf* cachet. This results in 15 different cachets of origin -- 3xBp + 3xMf + each of the other nine cities.

Flight 1, which took place on September 4, made the circuit Budapest–Pécs–Kaposvár and return to Budapest. The card shown in Figure 5–46 was carried on the entirety of Flight 1 and shows the transit marks of both Pécs and Kaposvár. The Flight 1 cachets I have seen are blue, except for the *Mf* origin, which is purple.

Flight 2 made the circuit Budapest–Szeged–Békéscsaba–Budapest on September 11. Figure 5–47 shows a cover carried on the Szeged–Békéscsaba and Békéscsaba–Budapest legs of this flight. The Flight 2 cachet is purple for all origins.

Flights 3, 4, and 5 together represented a multi–day circuit beginning and ending in Budapest, with one intermediate flyover of the airport at Mátyásföld. On September 16, Flight 3 went

Figure 5-48 Drop Mail Carried on Flights 3 and 4

from Budapest to Székesfehérvár and then on to Szombathely. On the 17th it continued as Flight 4 traveling Szombathely-Debrecen-Nyiregyháza, with drop mail to the Mátyásföld airport, near Budapest, enroute. Finally, Flight 5 flew from Nyiregyháza to Miskolc and then back to Budapest on September 18.

Some collectors look for all the possible routes. The total number of different covers is quite large. For instance for Flight 1 alone it can be nine pieces. Because mail was carried over all combined parts of Flights 3, 4, and 5, the combinations become very large, especially when the drop mail destination is included. Hózer listed 56 "different" Star Flight covers, and believed others existed.

Figure 5-48 shows a card carried on Flight 3 and part of Flight 4. It was dropped by parachute to Mátyásföld, and has an unusual large variation of the Star Flights vignette. This card originated at Mátyásföld.

The Flight 3 vignette is light green for all origins, and it is often difficult to see. In the figure it is at the lower center, just above the Mátyásföld receiving mark. The Flight 4 cachet, at the upper right in Figure 5-48, is purple for both origins. The drop mail, such as this, also shows a red or purple parachute-shaped handstamp indicating it was dropped to the Mátyásföld airport. In the figure, it is seen at the lower left, imprinted on top of the vignette.

Figure 4-49 shows a card carried on the final leg of Flight 5, from Miskolc to Budapest. The Flight 5 cachet is lilac from Nyiregyháza and purple from Miskolc. This card also has a vignette prepared for a domestic flight event earlier in the year, overprinted for the Star Flight.

On October 2 an additional flight Budapest-Kecskemét-Budapest was made by the Italian "Justice for Hungary." The vignettes and cachets were different, but again there were origin cachets for both Budapest and Mátyásföld, with a returning cachet from Kecskemét. It is notable that the stamp on much of the Kecskemét-Budapest mail was canceled by the violet cachet rather then the normal post office circular date stamp. Figure 5-50 shows this usage.

Figure 5-49 Card Carried on Star Flight 5

Figure 5–50 Kecskemét–Budapest Flight Card

Figure 5–51 Endresz Memorial Flight Drop Mail

The final *Justice for Hungary* airmail commemoration was a Budapest–Vienna *György Endresz Memorial Flight* that was scheduled for October 24; but, it was actually flown on October 31.

This flight was made in conjunction with an aviation exhibition that took place in Budapest from October 15–24. Because of the overwhelming crowds, the exhibit was extended to November 2 and the flight was delayed a week. The stamps on this *George Endresz Memorial Flight* mail are canceled with a special two–ring *Budapest 72 / Endresz Aviatikai Kiallitás* postmark.

From this flight, 404 postcards were dropped enroute at Györ. The remaining 288 cards were carried on to Vienna. Figure 5–51 shows an example of the Györ drop mail. The triangular flight cachet is blue, and the circular parachute drop hand–stamp is pink.

The Gerle Circum–Mediterranean Flight

A recent 10 Ft Hungarian airmail stamp pictures a 1930's biplane named the *Gerle 13.* That name probably draws a blank with many readers. Yet the saga of this airplane -- and its circum–Mediterranean flight -- made headlines in Hungary and throughout the aerophilatelic world in 1933.

The *Gerle 13* was a tiny sport biplane, with a wingspan of less than 30 feet. The pilots that made this 12,258 km trip

Figure 5–52 The Gerle 13

were Tibor Bisits and Antal Bánhidy. Figure 5-52 shows a photo of the plane, signed by the pilots. They were inspired by the *Justice for Hungary* flight from the U.S. to Hungary. But, that flight was in an American-built airplane. Bisits and Bánhidy wanted to make a record-breaking flight in a plane built in Hungary.

Figure 5-54 Gerle 13 Card Flown Oran-Budapest

The *Gerle 13* was designed by Bánhidy, and built by the Sport Flying Association of the Technical University of Budapest. Two such planes were built, and they were closely modeled after the *W.M.10* in production since 1931 by the Manfred Weiss Works in Budapest.

The *Gerle 13* bore the registration number HA-AAI, and was powered by a 100hp Genet-Major engine built by Armstrong Siddeley in England. In addition to its best-known flight around the Mediterranean, the *Gerle 13* was taken on an extended flight to northern Europe and Britain in 1933. But philatelicly, it is the flight around the Mediterranean that is of principal interest.

Takeoff from Budapest was on February 19, 1933. They flew counter-clockwise around the Mediterranean in 34 days, traveling from Hungary through Austria, Italy, France, Spain, Algeria, Libya, Egypt, Palestine, Syria, Turkey, Greece, Italy, Austria, and returning to Hungary on March 24. Mail was carried to, and picked up from, only five cities: Oran, Algeria; Tripoli, Libya; Jerusalem, Palestine; Athens, Greece; and Rome, Italy. [Actually, the Palestine landing took place in Ghaza.] The most common collecting approach is to collect mail between Budapest and each of those five cities, in both directions. Figures 5-53 and 5-54 show such a pair of cards flown Budapest to Oran and Oran to Budapest. The latter is signed by both pilots.

Mail was dispatched from Budapest to each of the five cities. The special postage rate was 72 fillér for postcards, but was sometimes overfranked. Mail from Budapest usually bears a black, blue, and white vignette showing the plane and illustrating its route. It is inscribed "First Hungarian Mediterranean Circular Flight" in English and Hungarian.

Figure 5-53 Budapest-Oran Card of the Gerle 13

Figure 5-55 Rome–Budapest Gerle 13 Flight Card

Also, a diamond–shaped bright pink handstamp was applied to all mail dispatched from Budapest. It shows a sunrise over the seashore, and is inscribed with that same text as the vignette, but in Hungarian and French.

The table at the right shows the dates of dispatch and receipt of *Gerle 13* mail to each of these cities. [Note that dispatch postmarks may precede the date of departure.]

Mail was dispatched for return to Budapest from each of the five cities. This mail some–times bears the same vignette as the mail from Hungary -- and, of course, the appropriate postage for country of origin. There was no mail service between the stops enroute. On receipt in Budapest, the foreign mail had the standard double–ring Budapest airmail circular date stamp of March 24 applied, as a receiving mark. Also applied was a special light green handstamp. This handstamp is seen in Figures 5–54 and 5–55. Figure 5–55 is a card flown on the last major leg of the flight, from Rome to Budapest. This receiving handstamp is often very weakly applied and/or faded.

The scarcity of this mail is about equal for each of the five cities on the route. But, there is considerably more demand for mail between Budapest and Jerusalem due to the many Holy Land collectors. As a result, these bring considerably higher prices. Figure 5–56 shows a card flown from Budapest to Jerusalem on the *Gerle 13*.

One other interesting aspect of mail intended for the flight of the *Gerle 13* is mail apparently forwarded from Hungary that missed the return flight. Most of this mail was flown back to Hungary -- but by

City	Landed	Departed
Budapest		19 Feb
Oran	28 Feb	01 Mar
Tripoli	03 Mar	05 Mar
Jerusalem	10 Mar	12 Mar
Athens	17 Mar	18 Mar
Rome	19 Mar	21 Mar
Budapest	24 Mar	

Figure 5–56 Budapest–Jerusalem Gerle Flight Card

Figure 5-57 "Delayed" Mail Intended for the Gerle 13 from Jerusalem

normal airmail routes. It is typically handstamped with a marking that reads "Dispatched after the departure of the plane Gerle 13." That is the actual English text of the marking applied in Palestine shown in Figure 5-57.

From other cities, the equivalent handstamp is in an appropriate language(s). For instance, the Greek handstamp is in Greek and French. Note that some of this delayed mail "missed" the flight by several months. Dispatches even from May and June are known! While these "delayed" pieces are interesting, they are much less desirable than mail actually flown by the *Gerle 13* itself.

Fourth Boy Scout Jamboree at Gödöllö

The *Fourth World Boy Scout Jamboree* brought 21,000 scouts from 48 countries and colonies to the former royal hunting preserve at Gödöllö, just outside of Budapest. Flying, especially soaring, was a Boy Scout passion in that era. So, it was natural that a temporary airmail route would be established from the Jamboree site.

The *Boy Scout Airmail Service* made daily flights over the ten mile route from Gödöllö to Mátyásföld airport. It carried airmail to and from the scouts encamped there, including international mail that was flown to or from Budapest. Figure 5-58 shows the reverse of a postcard flown by the *Boy Scout Airmail*

Figure 5-58 Opening Ceremonies at the Fourth World Boy Scout Jamboree

Service. It shows Hungarian Regent Nicholas Horthy with Lord Baden–Powell, the founder of the scouting movement. They are pictured at the opening ceremonics of the Jamboree.

These Jamboree airmail flights were operated by four pilots, led by Frigyes Uefty who had organized the *Flying Scouts of Hungary*. According to Short, some of the mail "was reportedly flown by glider." This is based on the recollections of Uefty's son, a glider enthusiast himself. But,

Figure 5-60 Budapest-Gödöllö First Flight Card

Otto Schaffling, who was an avid philatelist and the official photographer at the Jamboree, told me he would have known if mail was carried by glider, and that "it was not."

The *Boy Scout Airmail Service* used four handstamped cachets on its mail. Circular cachets picturing an airplane and a stag were used for the first flights of August 1. The Gödöllö–Budapest cachet is light green, and the airplane is facing right. Figure 5–59 shows an example of a first–day airmail card flown from Gödöllö to Budapest, bearing one of the many

Figure 5-61 Gödöllö-Budapest Scout Airmail

vignettes, with both scouting and aviation themes, used at the Jamboree.

Figure 5-59 Gödöllö-Budapest First Flight Card

The first–day Budapest–Gödöllö cachet is colored pale yellow, and the airplane faces to the left. Both this and the green first–day handstamp are in very pale inks, and sometimes are difficult to see clearly. Figure 5–60 shows an example flown to Gödöllö.

During the rest of the Jamboree, from August 2–16, a semi–circular cachet picturing a fleur–de–lis was used. For the Gödöllö–Budapest direction this cachet is brown, with the fleur-de-lis at the bottom. An example is shown in Figure 5-61. This card shows the complete set of stamps issued to commemorate the Jamboree. Note also the "X" overprint obliterating the airmail etiquette, that was applied at Mátyásföld airport. This indicates that no airmail service beyond that point has been prepaid.

The cachet for Budapest-Gödöllö is blue, with the fleur-de-lis at the top. Figure 5-62 shows an example of this last cachet. The mail of the 4th Boy Scout World Jamboree at Gödöllö exists in many minor variations, and is prized by specialists in Boy Scout philately.

For more information on other aspects of Fourth World Boy Scout Jamboree philately, see the article in Volume XVI, No. 1 of the journal of the *Scouts on Stamps International* society, by Sheldon S. Levy. Similar data can be found in Levy's series in *Linn's Weekly Stamp News* beginning June 2, 1969.

Figure 5-62 Budapest-Gödöllö Scout Airmail

Trans-Atlantic Airplane Mail of the 1930's

This chapter will deal primarily with the airmail services operated by *Air France* and *Luft Hansa* between Europe and South America during the 1930's. I will also discuss here the catapult system used to speed mail carried aboard certain German ships on the North Atlantic passages.

Figure 5-63 illustrates the French and German airmail routes to South America

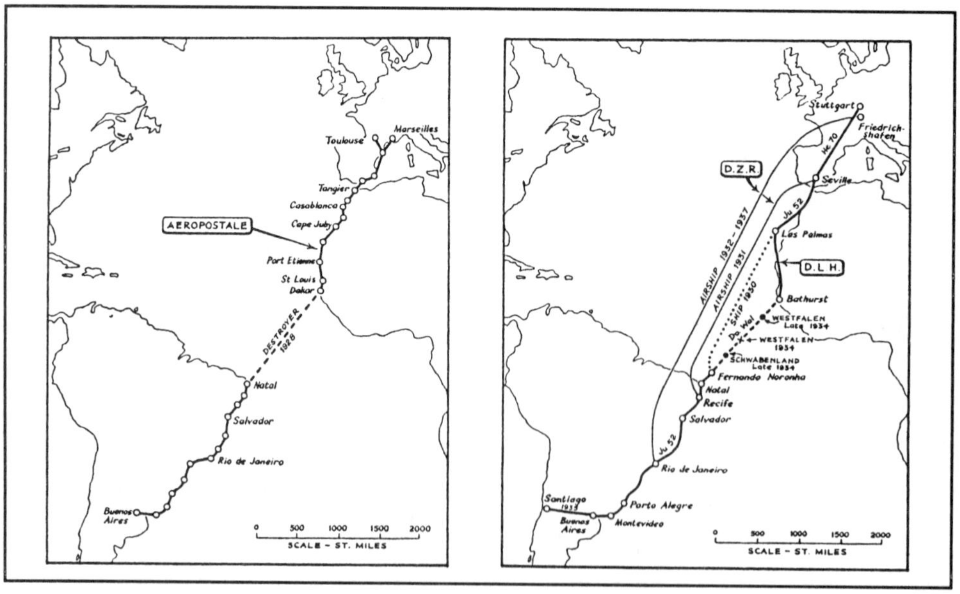

Figure 5-63 French and German Airmail Routes to South America

during the 1930's. These maps were adapted from Davies' *A History of the World's Airlines*. These services are discussed below.

Aéropostale and Air France Service. A French airline named *Aéropostale* began air-mail service to Morocco in 1919. In 1924, a French banker with major interests in Argentina began to take control of the airline and moved it in the direction of airmail service to South America. By 1925 its routes reached Dakar, Senegal at the westernmost point of Africa.

Figure 5-65 Hungary-Brazil Airmail of 1933

In 1927 the airline began a route in South America from Buenos Aires to Natal, near Brazil's easternmost point. On March 1, 1928, *Aéropostale* began mail service between Toulouse and Buenos Aires. This combined air and sea journey took eight days. Five of those days were for the journey by an out-dated destroyer from St. Louis, Senegal to Natal.

In 1930, the sea journey was first made by a flying boat that flew from Dakar to Natal in 19½ hours, cutting the full journey to four days. By this time, though, *Aéropostale* was having serious financial difficulties, and the all-air service was very sporadic.

Figures 5-64 and 5-65 show Brazil-Hungary airmail of 1932 and 1933 respectively, carried by *Aéropostale*. Both pieces had a nine-day transit time, and so were clearly carried by ship over the trans-Atlantic passage.

Because of its bankruptcy and the importance of its South America route, *Aéropostale* was included in the 1933 merger that formed *Air France*. By 1934, they finally had flying boats available that could make the non-stop trans-Atlantic journey reliably. The sporadic all-air service improved considerably, but it was not until January 1936 that a truly regular all-air service to South America was flown on a scheduled basis by *Air France*.

Figure 5-64 Brazil-Hungary Airmail of 1932

Figure 5-66 shows a 1936 airmail letter from Hungary to Rio de Janeiro carried by Air France on its all-air service. The letter is backstamped in Paris on May 9, and in Rio on May 12. Total travel time from Budapest was four days. Note that it bears a mixed franking of the 1933 and 1936 airmail issues.

Figure 5–66 Budapest–Rio Airmail via *Air France* in 1936

Deutsche Luft Hansa Service. In 1930, *Luft Hansa* began service to South America by flying airmail via Spain to Las Palmas, Canary Islands. There, the mail was put aboard ship to Fernando de Noronha off the coast of Brazil. *DLH*'s South American subsidiary, *Kondor*, flew the mail from that island to Rio de Janeiro.

As previously noted, in September 1931 the Graf Zeppelin began regular, seasonal bi–weekly passenger and airmail service to Brazil.

While the Zeppelin service went on, *Luft Hansa* began innovative experiments in 1933 using a depot ship -- the *Westfalen* -- anchored in mid–Atlantic. The idea was to pick up incoming seaplanes, refuel them, and then launch them by catapult. This DLH route was from Bathurst, Gambia to Natal.

On February 3, 1934 *Luft Hansa* inaugurated the first regular trans–Atlantic airmail service on this route with a westbound flight. Schedule time from Berlin to Buenos Aires was four days. Hungarian mail carried on that flight is known, but it is quite rare.

Figure 5–67 shows an interesting postcard that ties together the French and German airmail services to South America. It is a January 25, 1934 card, from a well–known Hungarian collector, to Brazil. It asks the recipient to send him ten postcards using the first eastbound *DLH* flight via the *Westfalen*, that was scheduled for February 7. Unfortunately, the sporadic *Air France* all–air service wasn't running that day, and the card did not arrive until February 10. This, perhaps, tells us why no mail to Hungary is reported from the first eastbound flight!

DLH was experiencing difficulties attaining maximum useful load at takeoff due to the tropical heat. [The lower density of very warm air results in less lift being produced by aircraft wings.] Therefore, later in 1934, they moved the *Westfalen* to the coast of Gambia, and placed a second ship -- the *Schwabenland* --

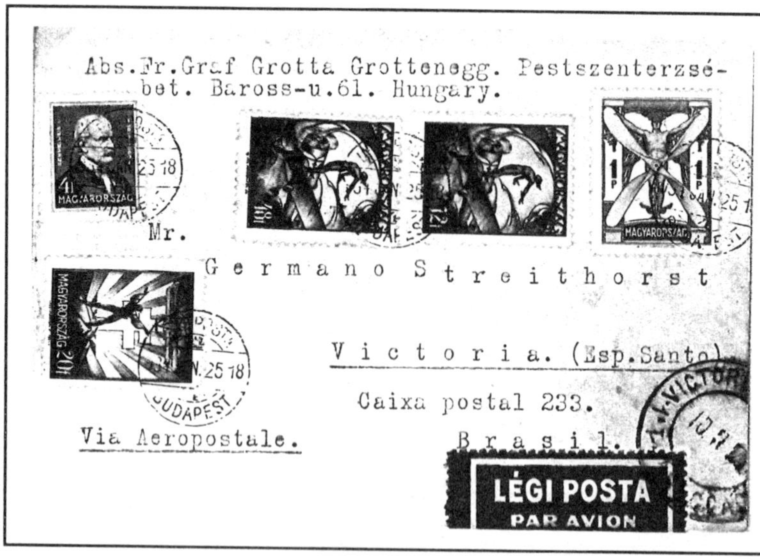

Figure 5–67 Air France Mail Relating to DLH First Flight

near Fernando de Noronha. This increased schedule reliability by reducing the necessary fuel load the seaplanes had to carry. At times, the ship carried the airplane away from the African coast to get cooler temperatures for launching.

Figure 5-68 DLH 1st Anniversary Flight to SA

Figure 5–68 shows a 1935 letter from Hungary to Brazil. It bears the first anniversary cachet of the German airmail service to South America. Figure 5–69 shows 1937 mail from Hungary to Brazil flown by *DLH*. Note the red, double-ring *Deutsche Luftpost / Europa–Südamerika* handstamp that was often -- but not always -- applied to mail carried by *DLH* to South America. Haberer's *Part 2* catalog provides detailed information about the dates of all *DLH* flights on this route. These flights continued, using depot ships for refueling, until the beginning of World War II.

Friendly rivalry had characterized the airline industry from its beginnings. On the South Atlantic route, this became true cooperation between *DLH* and *Air France*. It became apparent to both competitors that the integration of services permitted by "pooling" saved considerable cost. These pooling arrangements, with a weekly flight in each direction by each airline, continued until the beginning of the war.

Figure 5–70 shows 1937 airmail from Hungary to Uruguay carried by *Air France* with a *Le Bourget / Port-Aerien* transit mark.

The principal means for distinguishing between mail carried by *Air France* and mail carried by *Lufthansa* are Paris or Berlin transit marks, or the red *Luft Hansa* handstamp seen in Figure 5–69. [Note: Zeppelin flight dates given by Sieger and *Luft Hansa* flight dates given by Haberer *may* be required to distinguish whether a piece of mail with this marking was carried by Zeppelin or by airplane.] Remember, the mailer's instructions and the use of specific-language airmail etiquettes did not determine the airline used, as can be seen from Figure 5–70. This was a genuine pooling arrangement, in

Figure 5-69 Hungary-Brazil Airmail of 1937 Flown by DLH

Figure 5-70 Hungary-Uruguay Mail Flown by Air France

the model that became universal in the post-war years. Mail was sent by the most expeditious flight.

By 1939, it often becomes difficult to determine the airline that carried a piece of mail. For instance, Figure 5-71 is a letter from Hungary to the unusual destination of Trinidad in the British West Indies. There are no intermediate transit marks. A typed instruction of "via Natal" implies air service to South America (rather than the sea route via North America).

Figure 5-72 *Nachbringe* Flight to the *Columbus*

The six-day journey from Budapest to Port of Spain, documented by postmarks, proves the letter was flown. Since the letter left Budapest's airport on February 4 and arrived in Trinidad on February 10, it could not have been on either *DLH*'s February 2nd or 9th flight out of Berlin. Therefore, it must have been flown trans-Atlantic by *Air France*. Which airline(s) carried it from Brazil to Trinidad remains a mystery.

Figure 5-71 Hungary-Trinidad Airmail Flown by Air France

North Atlantic Catapult Mail. All-air service across the north Atlantic did not begin until May 1939. In earlier years, special arrangements were made to speed airmail to its destination, although most of the north Atlantic passage was aboard ship. Three German liners were involved in these arrangements: the *Columbus*, the *Bremen*, and the *Europa*.

The first part of the arrangement brought airmail to the ships after they departed Germany. Special flights were made with airmail from Cologne to Cherbourg, France where the ships put in to port. These so-called *nachbringe* flights saved up to two days in transit time. Figure 5–72 shows a very scarce example of 1932 Hungarian mail flown on a *nachbringe* flight to the *Columbus*.

Figure 5–73 Catapult Airmail from the *Europa*

In addition to *nachbringe* flights, the *Europa* and the *Bremen* -- but, not the *Columbus* -- had catapults aboard which launched small seaplanes that carried mail from ship to shore. By launching the plane 600–700 miles from New York, another day–and–a–half could be saved in airmail transit time.

The *Bremen* began ship–to–shore mail in 1929 and the *Europa* in 1930. This type of operation continued seasonally until the end of 1935. The Zeppelin *Hindenburg*'s North American route which began in 1936 eliminated the need for continuing the catapult service. Haberer's *Part 1* catalog lists all these catapult mail flights. Mail was carried on these catapulted aircraft only based upon a contract -- similar to the "treaty states" contracts discussed under Zeppelin mail. Occasionally, mail from non–treaty states was flown. Hungary's status as a treaty state became effective on August 18, 1931. Hungarian mail is known from only a single flight before that date, October 16, 1930.

In all, Haberer lists 41 ship–to–shore flights from either the *Europa* or the *Bremen* that carried Hungarian mail. The most Hungarian mail carried on any flight was 17 items; but, the average was only seven. This mail is obviously very scarce, in great demand, and so quite expensive.

Figure 5–73 is a card flown from the Europa to New York on March 25, 1932. Figure 5–74 is a card flown from the Bremen to New York on September 9, 1934.

There was also American mail catapulted ship–to–shore on the return voyages. The destination of the catapulted airplanes in this case was usually Southampton, England. In the early years, they occasionally flew to

Figure 5–74 Catapult Airmail from the *Bremen*

Amsterdam or Cologne. Eastbound catapulted mail addressed for Hungary may also be of interest to Hungarian aerophilatelists.

A similar catapult arrangement was used experimentally by the French liner *Ile de France* from 1928 to 1930. No Hungarian mail is reported to have been flown from that ship.

Other 1930's International Airmail

We have seen the beginnings of airmail to the Americas by both airship and airplane, as well as some special events in Hungarian airmail history. In the 1930's, airmail was well developed throughout Europe, and it was rapidly progressing in Africa and Asia. This chapter will focus on this "other" international airmail of the period.

Because it is so open-ended, this material is extremely difficult to organize, both from the perspective of a collector and of a writer. Therefore, I have been arbitrary. I will deal with the following topics: the airmail stamps of the 1930's; new European airline services to Budapest; airmail to Africa and Asia; and finally I will mention and illustrate a "mixed-bag" of interesting airmail items that are hard to categorize.

Airmail Stamps of the 1930's. In 1930, four values were added to the airmail issue of 1927. These airmail stamps were valid until December 31, 1933. On April 20, 1933 a new set of airmail stamps was issued. This set of nine stamps consisted of four designs. As a result, for over eight months mixed frankings of the two issues were possible. These make interesting additions to a collection.

The two low values of the 1933 issue, picturing the *Justice for Hungary*, were designed by Ferenc Helbing. The other values showing the *Giustizia per L'Ungheria*, an allegory of flight, and Mercury with a propeller were designed by Ferenc Márton. The total number of 5 Pengő stamps sold was 11,812 making it by far the scarcest of all Hungary's normal airmail stamps. The issue exists imperforate.

An interesting fact about these stamps was that the four low values were not permitted entry into Czechoslovakia due to their "political" motifs. Therefore, mail to Czechoslovakia franked with these stamps was refused and returned to the sender.

Figure 5-75 shows the entire issue of 1933 used on a special flight cover to the famous *WIPA* stamp exhibition in Vienna. The Hungarian *WIPA* flight mail usually comes in two forms ... the six low values on a postcard that is quite inexpensive, and the entire set on cover that is costly.

Figure 5-75 The Issue of 1933 on WIPA Flight Cover

Actually there were three *WIPA* flights. The special green cachet shown was applied only to the first on June 24. There is also a special blue cachet for Austrian mail carried on the first *WIPA* flight to Budapest. The other *WIPA* flight covers are of June 30 and July 5. They show only the *WIPA Wien* receiving mark.

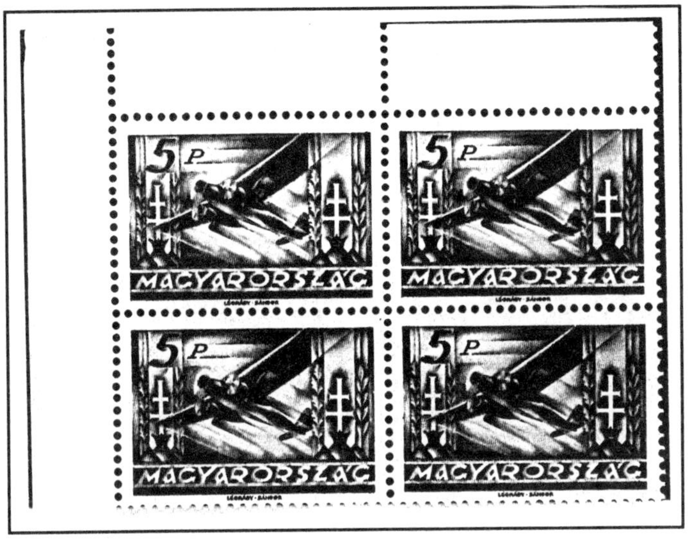

Figure 5–76 High Value of the Issue of 1936

The airmail issue of 1936 was released on May 8th. It was valid until March 31, 1939. Figure 5–76 shows the 5 Pengő value of this issue. The award–winning designs were by Sándor Legrády. This issue exists imperforate.

Again, the overlap in validity between the issues of 1933 and 1936 was about eight months. Mixed frankings of these two issues are interesting to seek out.

Figure 5–77 shows a 1937 cover to Austria franked with the entire issue. When this issue became invalid in 1939, Hungary had no airmail stamps for eight years.

Another infrequently seen set of "stamps" are six 1933 semi-officials by *MALÉRT* intended for air parcel post. Though found both mint and canceled with a company handstamp, these stamps were never put into use.

Airlines Servicing Budapest. Through the 1930's several airlines extended their European services to Budapest. [These are in addition to the intercontinental services operated by *IA*, *DLH* and *KLM*.]

The first of these services was *Transadriatica Airlines* that provided a short–lived through service from Budapest to Venice. The daily flights began on August 15, 1931, leaving Budapest at 9 AM and arriving in Venice at 3:30 PM. First flight mail carried on this route bears a pink cachet.

Though not a new route, an improved service between Paris and Budapest was instituted in August 1933, when *CIDNA* was among the companies merged to form *Air France*. A special card with both French and Hungarian postage demonstrated that the Budapest–Paris–Budapest round–trip could be completed in "less than 25 hours."

Figure 5–77 The Issue of 1936 on Cover to Austria

Imperial Airways added Budapest as the terminus of its European service on April 1, 1935. [Note: this is a different route than the IA intercontinental route that passed through Budapest briefly in 1929, and then again in 1930–31.] This new service ended in 1937.

The London–Cologne–Budapest route made a stop in Brussels only initially. Figure 5–78 shows a very unusual Belgian first flight cover to Budapest, signed by the pilot. Figure 5–79 is a westbound first flight cover of April 2. Note the unusual *IA* receiving mark.

Figure 5–78 Belgian FFC on IA London–Budapest Route

Figure 5–79 Imperial Airways Budapest–London FFC

Also on April 1, 1935, another Italian airline came to Budapest. *Ala Littoria* inaugurated service between Budapest and Venice via Vienna. A special large circular receiving mark was prepared for the inbound first flight mail of April 1, and a similar cancellation for the April 2 first flight mail leaving Hungary for Italy.

On April 20, 1936 *KLM* extended its European service from Amsterdam to Budapest via Prague and Vienna. Cacheted first flight covers with Budapest arrival cancels of April 20 and Budapest dispatches of April 21 are available. From this time, until the beginning of the war, Budapest was on *both KLM*'s European and inter-continental services. From 1938, the western terminus of *KLM*'s European service to Budapest was London, and it was routed via Rotterdam rather than Amsterdam.

Sabena extended its European operations from Brussels to Prague on April 5, 1937. First flight mail carried on this route -- using other airlines for the connection between Budapest and Prague -- is available.

On May 3, 1938, a Budapest–Arad–Bucharest service was initiated. This route was jointly operated by *MALÉRT* and the Romanian national airline *LARES*. First flight covers are often canceled at the Budapest International Fair and bear vignettes issued for the beginning of the service.

On July 1, 1938 a new direct service jointly operated by *MALÉRT* and *PLL*, the Polish national airline, was instituted between Budapest and Warsaw. Only a violet Polish *Warszawa–Budapest* handstamped cachet was prepared for this inaugural. Figure

Figure 5-80 First Direct Flight Warsaw-Budapest

5–80 shows Polish mail to Budapest flown on that first flight. Hungarian mail, flown in both directions, with the Polish cachet, exists. Apparently, the *PLL* participation in this route was very short–lived, since they are not mentioned in either the 1938 or 1939 edition of Jane's. But, the route continued to be flown by *MALÉRT*.

On August 24, 1938 *MALÉRT* began operations between Budapest and Prague. At the same time, *CSA* (the Czechoslovak national airline) began operations between Budapest and Bratislava. Hungarian mail of these first flights was prepared in conjunction with the *8th FILPROK* stamp exhibit. Therefore, they are rather gaudy, typically with stamps, meters, cachets, vignettes, and several postal markings.

In addition to its new routes later in the decade, *MALÉRT* also made two other route changes earlier. From 1932–35 they flew a twice weekly route from Budapest to Venice via Klagenfurt, Austria. Latham lists a June 16, 1932 first flight of this route. Beginning in 1936, *MALÉRT* extended its original Budapest route to Zurich, via Salzburg, Austria and Munich.

Below I have summarized the status of airline operations to and through Budapest in the summer of 1939, just prior to the beginning of the war. This information is based on the 1939 edition of Jane's *All the World's Aircraft*.

Air France: Budapest was the stop between Vienna and Belgrade on their Paris–Bucharest service.

Ala Littoria: Budapest–Vienna–Venice–Rome.

CSA: Budapest–Bratislava.

DLH: Budapest was the stop between Vienna and Belgrade on Berlin–Athens European service; Budapest was the stop between Berlin and Belgrade on the Berlin–Teheran intercontinental service; also *DLH* operated a direct Berlin–Budapest–Bucharest route. With the 1938 *Anschluss*, *ÖLAG*'s routes were merged into *DLH* services.

KLM: Budapest was the terminus of London–Budapest European service; it also was the stop between Leipzig and Athens on *KLM*'s Amsterdam–Batavia intercontinental service.

LARES: Budapest–Arad–Bucharest.

MALÉRT: Budapest–Vienna–Salzburg–Munich–Zurich; Budapest–Arad–Bucharest; Budapest–Warsaw; and Budapest–Prague.

With the beginning of the war, most international air routes ceased operation. The only new route to Hungary was a *MALÉRT* pooling operation with *Aeroput*, the Yugoslav national airline, which began a short–lived Budapest–Belgrade service in 1940. I have never heard of first flight covers for this route.

For a short time, *MALÉRT* continued operations to Romania. As will be noted, their domestic operations ceased in January 1941. Until 1943, *MALÉRT* continued operating between Budapest and Vienna, in pooling operations with the German and Italian airlines. But, all such operations were tightly controlled by *DLH* and the German military.

Figure 5-81 Budapest-Cairo 1930 FFC via *IA* Service

Airmail to Africa and Asia. As previously noted, *Imperial Airways* made two flights via Budapest in late 1929 on their all-air London-Cairo service. From April 1930 to May 1931, the European routing was again via Budapest. During this period, a short rail segment between Skopje and Salonika was used.

Some very interesting airmail is available that was carried on the *IA* route via Budapest during its 13 months of operation. Figure 5-81 shows a letter flown to Cairo on the first flight of that route.

The first major African segment of *IA*'s *London-Cairo-Cape* airline route began operation in February 1931 to Mwanza, Tanganyika. Figure 5-82 shows a Budapest-Mwanza first flight cover of that route.

In May 1931, *Imperial Airways* went back to a rail connection from Basle to Genoa, and by December the railroad was used from Basle all the way to Brindisi, Italy. From these ports, flying boats carried passengers and mail across the Mediterranean to Egypt.

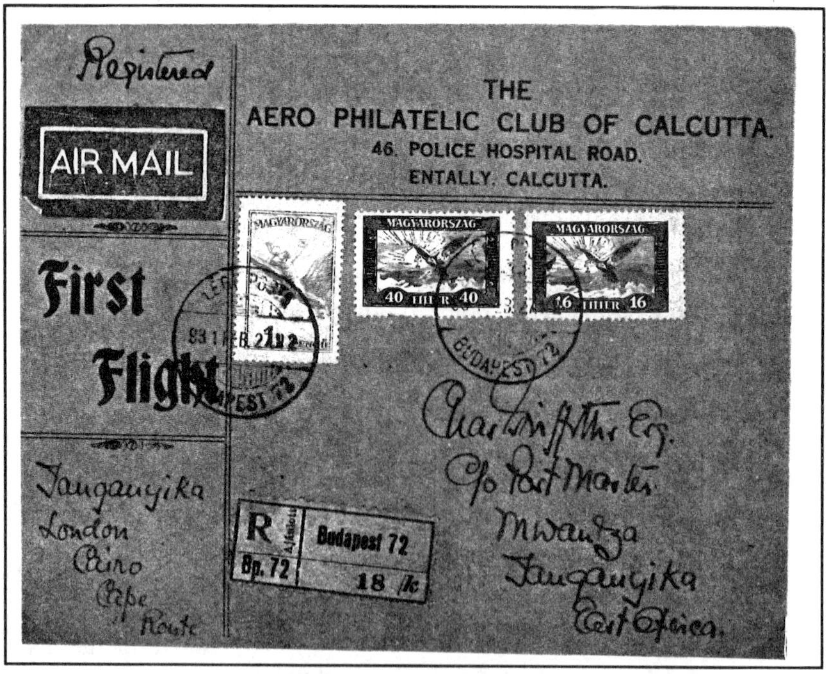

Figure 5-82 Budapest-Mwanza 1931 FFC via *IA*

By 1929, *IA* was operating from Egypt to the west coast of India. Regular airmail service across India (Karachi to Calcutta) began on July 7, 1933. Hungarian mail carried on that flight exists.

Later, *Imperial Airways* services were extended to Singapore and then to Australia. During this period, Hungary was not on the *IA* intercontinental route. So, it was more likely that Hungarian airmail destined for Asia was flown by *KLM*, which operated its route from Amsterdam to Batavia via Budapest throughout the 1930's.

The initial thrust of *Deutsche Luft Hansa* eastward in the late 1920's and early 1930's was through the Soviet Union. Operating in conjunction with the *Russian State Airlines*, they reached China via Siberia by 1930. In 1930, *DLH* also organized a German-controlled airline within China called *Eurasia*.

Figure 5-83 FFC of 1931 Shanghai-Siberia-Berlin to Hungary

Figure 5-83 shows a June 1931 first flight cover flown from Shanghai to Hungary via Siberia and Berlin.

Figure 5-84 shows a cover flown from Hungary to Shanghai in 1932. It bears postal markings of Berlin; Rostov, USSR; Harbin and Shanghai, China.

These Siberian operations were only sporadic, and continuing political troubles between Germany and the Soviet Union doomed this route to defeat. They apparently ended operations by the mid-1930's.

Later in the decade *DLH* was exploring a southern route through Asia. An attempt to establish a route from Baghdad to China via Teheran and Kabul failed in 1937. On the return from China by the survey flight, a forced landing was necessary and hostile tribesmen held the crew prisoner for four weeks. This discouraged further exploration of that approach. In the spring of 1938, *DLH* established a regular route to Teheran, with a later extension to Kabul, Afghanistan.

Ultimately, *DLH* chose an Asian route via India to Bangkok. This route was inaugurated during the summer of 1939. As happened earlier in the Soviet Union, *DLH* was soon unwelcome in the area.

Beginning in February-March 1936, through mail from China to Europe was possible by another route. This new route was facilitated by the connection of the *China National Aviation* routes in China to *Air France*'s south Asia route at Hanoi. The latter route went to Europe via Bangkok, India, and Baghdad.

The first flight cover in my collection left Shanghai on February 12. It did not arrive in Hungary until March 17, being delayed for over three weeks "by adverse weather."

On November 17, 1935, *Sabena Airlines* inaugurated through service from Brussels to Elizabethville in the Congo. Though this route did not pass through Budapest, Hungarian mail forwarded to Brussels for the first flight is known.

Figure 5-84 Hungarian Mail to China via Siberia

Other Interesting Airmail Items. Gliders and soaring were a popular "craze" in the 1930's. Therefore, many glider mail experiments were conducted. A couple of domestic glider events will be mentioned in the next chapter. Here, I will start with a few international glider flights.

A glider mail item of interest is Hungarian airmail flown on Robert Kronfeld's Vienna to Semmering glider flight of January 1933. 446 pieces of Hungarian mail were carried.

Figure 5-85 Kronfeld 4-Nation Glider Flight

Kronfeld's *4-Nation Glider Flight* of July 1933 wended its way through Austria, Hungary, Italy, and Switzerland. Mail carried on this flight is much more pertinent to Hungarian aerophilately than the Semmering flight, but is also much rarer. The flight made a circuit from Vienna via Budapest, Trieste (Udine), Milan, Zurich, Innsbruck, Salzburg and back to Vienna. Hungarian mail flown from Vienna to each destination is known, but quantities are very small. [For this material, by "small quantities" I mean about 100 pieces. In some cases, only five or six pieces were flown.]

Budapest was the only destination to produce a special receiving mark. This gold, rectangular handstamp is collected on the mail of all participating countries (plus Liechtenstein, whose mail was put aboard the glider at Zurich). Except for the 1,300 Austrian postcards flown to Budapest, this foreign mail with the Budapest receiving mark is also rare.

The item most in demand by Hungarian collectors is Hungarian mail flown on Kronfeld's glider from Vienna to Budapest. This mail bears both Hungarian airmail stamps and the Hungarian receiving mark. One of the 88 such pieces flown is shown in Figure 5-85.

A meeting of the *International Study Commission for Motorless Flight* (ISTUS) was held in Budapest from May 18-24, 1936. A flight to demonstrate the feasibility of glider trains was arranged by the Austrian Aero Club. An airplane towed two *Rhönsperber* gliders from Vienna to Budapest on May 19.

The "sky train" left Vienna's Aspern airport at 7AM with 3,551 pieces of mail, the tow-plane piloted by Hans Bruckner. Enroute, one glider,

Figure 5-86 "Sky Train" Autograph Card to Budapest

piloted by Ignaz Stiefsohn, dropped about 25% of the mail to Sopron. The other glider, piloted by Bruno Gumpert, dropped another 25% to Györ. The balance was delivered to Budapest. All the mail was marked with a violet cachet picturing the glider train going from Vienna's St. Stephen's Cathedral to Budapest's Parliament Building.

The drop mail is identifiable by a handstamp picturing a parachute, and an appropriate receiving mark. Included in the mail to Budapest were 140 of the 200 numbered autograph cards signed by all three pilots and the Austrian and Hungarian organizers of the event. Figure 5–86 shows an example of that autograph card flown to Budapest.

On a flight from Budapest during the conference, Gumpert made an off-field landing in Czechoslovakia. He was detained there too long to participate in the return flight.

The return flight, with only one glider in tow, was made from Budapest to Vienna on May 24. Stiefsohn carried the 876 pieces of mail in his glider. This mail bears a similar black cachet to that of the Vienna–Budapest mail; but, the glider train is pointing in the other direction.

In addition to the cachet, some of the mail has a black and white vignette of the *Hungarian Aero Club* or a green and white vignette of the *6th FILPROK* stamp exhibition. Figure 5–87 shows an example of the latter. The rarest of this mail are the 10 pieces with imperforate copies of the black and white vignette.

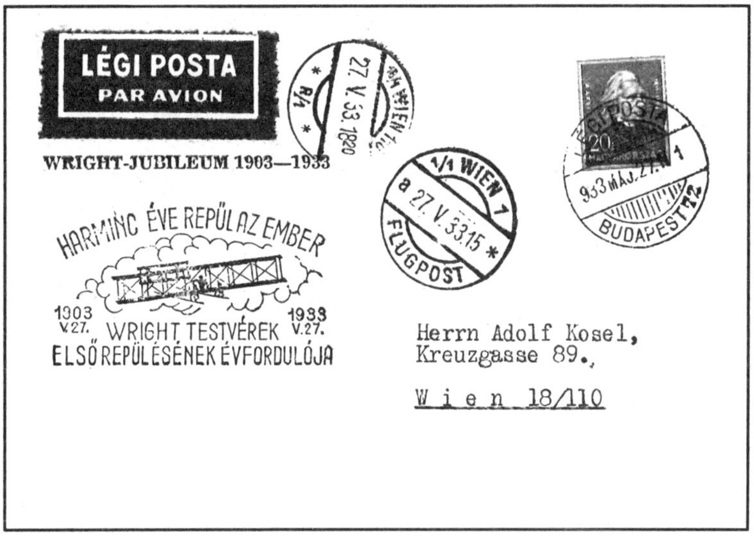

Figure 5–88 Wright Brothers Commemorative Cachet

I'll now show two items that are of particular interest to American collectors of Hungarian airmail. Figure 5–88 shows a 1933 Budapest–Vienna flight card with a black cachet commemorating the 30th anniversary of the Wright Brothers first manned flight. This was the first foreign philatelic commemoration for the Wrights. Though the cards do not indicate this, I believe this cachet was produced in conjunction with the *4th FIKPROK* stamp exhibit in Budapest.

The November 1936 airmail cover to the U.S. shown in Figure 5–89 seems unre–

Figure 5–87 Budapest–Vienna Glider Train Flight

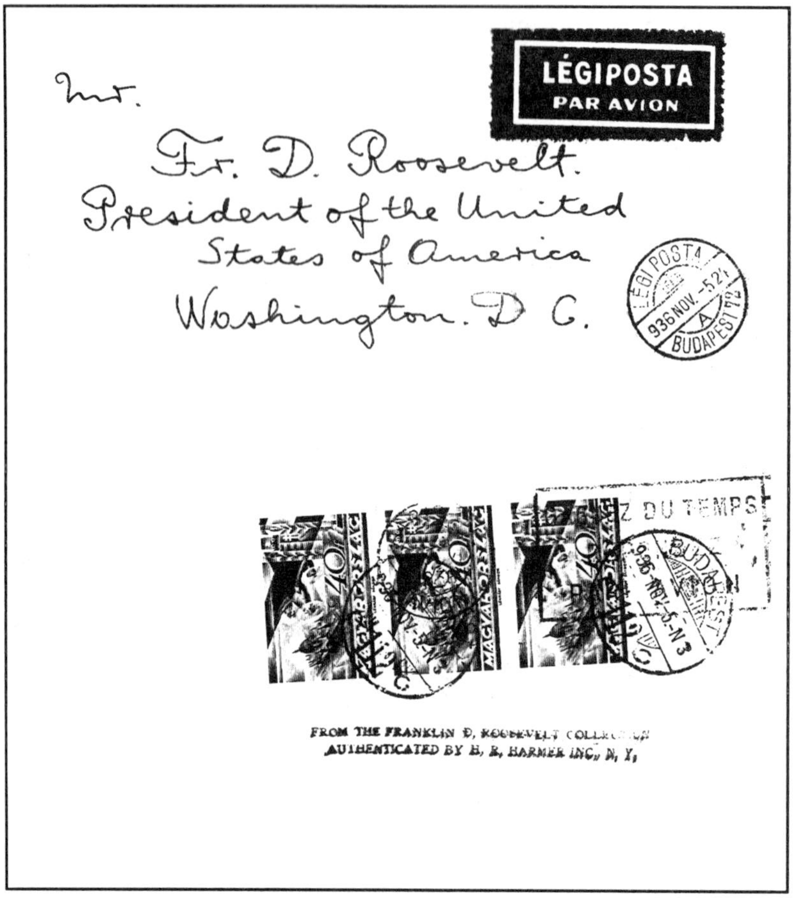

Figure 5-89 Hungarian Airmail to FDR

markable. It has a common franking, and, this being before trans-oceanic service, it was carried by air only as far as Paris.

What is interesting about it is that it's addressed to President Franklin D. Roosevelt, and was a part of his personal collection. It was another "fun" cover to find and add to my collection.

The giant 12-engine Dornier flying boat -- the *Do X* -- was supposed to make a flight from its base in Passau to Budapest a few months after its return from its historic trans-Atlantic flight. The flight was canceled due to repairs needed by the aircraft. Mail of May 12, 1933 with the flight cachet exists. The cachet is usually obliterated, and a 4-line violet handstamp announcing the cancellation of the flight was added.

One of the great heroes of Hungary was Ferenc Rákóczi II. In 1935, a set of regular stamps was issued to commemorate the 200th anniversary of his death. At that time a special flight was made from Budapest to Istanbul, near the town of Rodosto (now Tekirdag), where Rákóczi lived out his life in exile. Figure 5-90 shows a large cover carried on this flight with the cachet and vignettes produced for the event by the *5th FILPROK* stamp exhibit.

Figure 5-91 shows a postcard flown in May of 1936 on the Austrian Whitsunday Flight (*Pfingstflug*)

Figure 5-90 Rákóczi Anniversary Flight to Istanbul

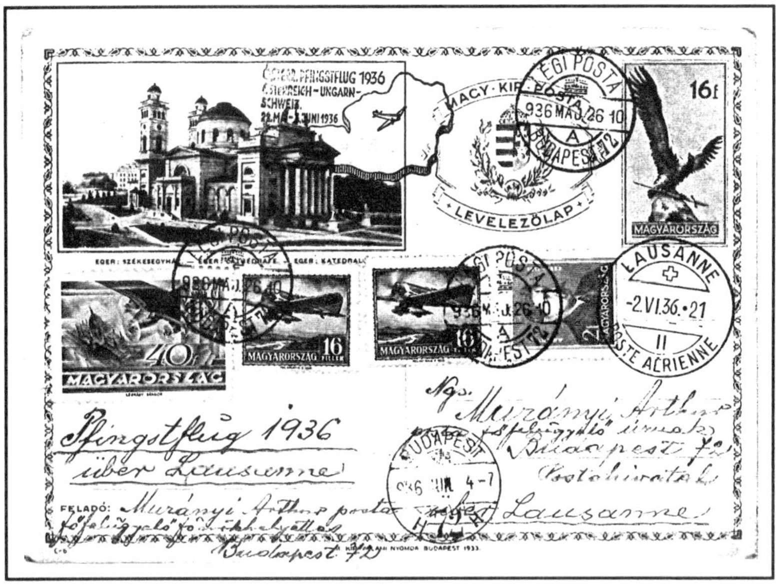

Figure 5-91 Austrian *Pfingstflug* Card with Mixed Franking

through Austria, Hungary, and Switzerland. This card was carried on that flight as far as Lausanne. It is also interesting because it shows a mixed franking of the Hungarian airmail issues of 1933 and 1936.

Finally, I'd like to show a card that draws attention in several different ways. Though this was their intended use, Hungarian international reply cards are only rarely seen with foreign postmarks. It is even more unusual for them to be flown back to Hungary by airmail. Figure 5-92 shows such a card of November 1937. It is remarkable for two other reasons as well. The original sender affixed additional Hungarian postage -- including an airmail stamp -- to cover the cost of airmail return. Also, the Hungarian postage is canceled by an anti-Semitic handstamped canceller of Munich. That it was actually flown, on its return to Hungary, is established by the Vienna airmail transit marking.

Figure 5-92 Airmail Reply Card from Munich

I feel that I have not done justice to the topic of these chapters. There are many ways that international airmails of the 1930's could be collected. I've tended to focus on events and special markings. This is really a focus on philatelic mail. Alternatively, collecting only commercial airmail between Hungary and countries around the world presents many interesting possibilities -- and it would look less "made to order." With the large quantity of 1930's material available, there is probably a collecting approach to suit the individual interests and pocketbook of every collector of Hungarian airmail.

Domestic Airmail Events of 1931-41

In the chapter on *Domestic Airmail Events of 1924-29*, it was mentioned that several domestic airline routes supposedly initiated in 1928 and 1929 show no evidence of airmail until the 1930's. These are the routes that we will discuss here first.

The first of the continuing domestic air routes from which airmail is seen from time-to-time is Budapest-Pécs-Kaposvár. A postcard in my collection is the basis for Latham's listing of April 15, 1930 as the inaugural of that route. The card is addressed to Pécs and bears a manuscript

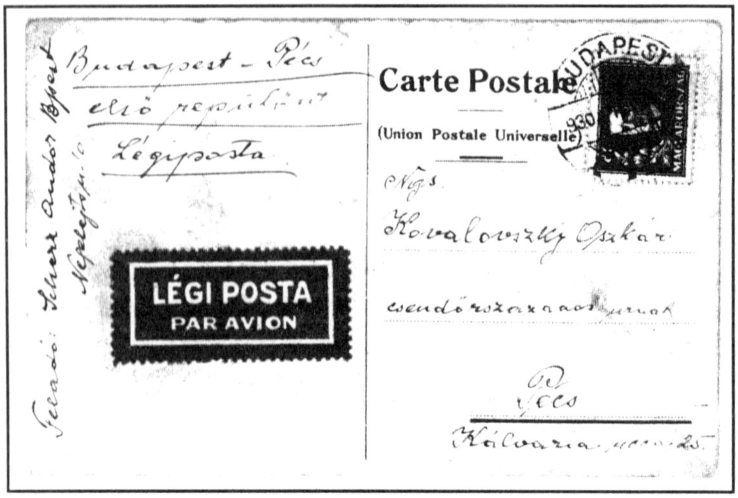

Figure 5–93 Budapest–Pécs First Flight Card

inscription about the first airmail journey between those cities. Figure 5–93 shows that Budapest–Pécs first flight card.

Jane's 1930 edition lists the following airfields in Hungary (with their primary purpose): Mátyásföld (airport of Budapest), Csepel (private field of the Weiss Works), Szombathely (training), Szeged (meteorology), Székesfehérvár (experimental flying), Kaposvár (airmail), Nyiregyháza (airmail), and Pécs (airmail).

In 1931, the principal domestic airmail events are the June 15 inaugural flights from Budapest to Nyiregyháza and return. Figure 5–94 shows a cover flown in the Budapest–Nyiregyháza direction. This cacheted mail is quite common, and the date is corroborated by Jane's 1932 edition that states that this air route operated from June 15 to September 19 in 1931. Jane's also notes that the Budapest–Pécs–Kaposvár route operated from May 11 through September 21. Latham's second addendum mentions a first flight cover of that date, but I have not seen such an item.

While Jane's in 1932 listed the total amount of mail carried on these routes in 1931, it is not meaningful to us since they included printed matter and parcels. In all probability, the bulk of the mail was Budapest newspapers being delivered to the provincial cities. Jane's also noted that in 1931 the government subsidy to *MALÉRT* was 200,000 Pengó.

In 1932, the Budapest–Pécs–Kaposvár route continued to be operated, but service between Budapest and Nyiregyháza was suspended awaiting construction of a new airport. From August 1 to September 1, *MALÉRT* operated an air route between Budapest and Miskolc. Cacheted airmail of both legs of this route dated August 1 are again quite common. Figure 5–95 shows a card carried in both directions on the first day of this route.

Beginning with 1932, Jane's lists the airfields in Miskolc and Debrecen as being the "landing grounds" of local flying clubs. After 1934, Jane's does not list any domestic air routes in Hungary until 1940. Apparently, *MALÉRT* suspended regular domestic operations. Throughout most of the rest of the 1930's domestic airmail in Hungary essentially means

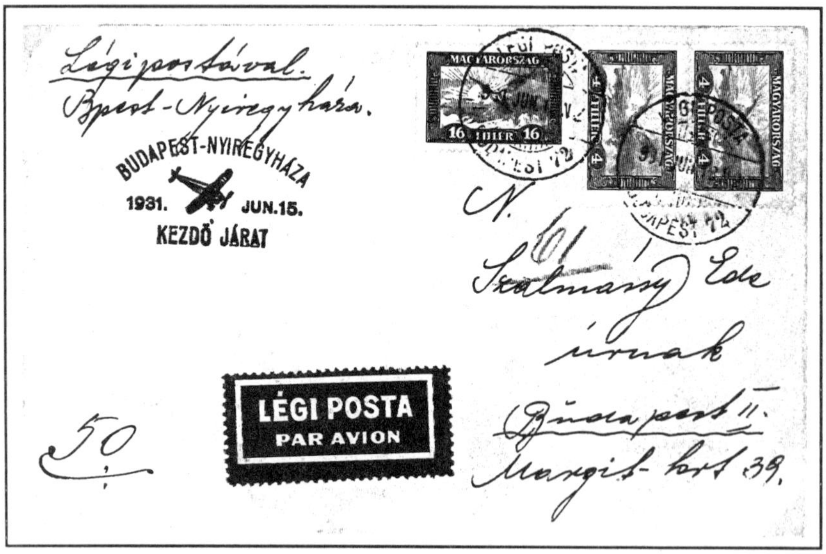

Figure 5–94 Budapest–Nyiregyháza First Flight Cover

Figure 5-95 First Flight of 1932 Budapest-Miskolc Route

"one-shot" flights associated with special events.

One trend concerning Hungarian aviation recorded by Jane's throughout the 1930's was the growing popularity of "soaring," i.e. flying in gliders. *The Hungarian National Gliding Association* (*MOVE*), whose president was István Petroczy, was first reported in 1930, and several sections were reported in provincial cities in later years. [Petroczy was a WWI Austro-Hungarian Army air corps officer, and with Theodore von Kármán built the *PKZ* helicopters in Budapest late in the war.] The popularity of soaring during the 1930's was worldwide. I will note both domestic and international airmail events involving gliders beginning in 1933.

The three major events of the 1931-33 period that bear on domestic airmail -- the *Justice for Hungary* Star Flights, the flight of the *Gerle 13*, and the *4th World Boy Scout Jamboree* airmail -- all were covered in separate chapters.

1933 had three minor domestic airmail events, and one of significant importance. In May, the *4th FIKPROK* exhibit was held in Budapest. [1933 was the first time *FIKPROK* included an airmail motif. As the decade went on, they were among the leaders -- with the *Miskolc Stamp Club* -- in producing aerophilatelic covers of rather poor taste.]

Their airmail event in 1933 was a flight of the balloon *Hungaria* piloted by a Major Poppe. On May 15, it carried mail from Budapest that was

Figure 5-96 Balloon Mail of the 4th FIKPROK Exhibit

dropped at Pestszenterzsébet. Four balloon vignettes were issued; so, four balloon cards each with a different vignette makes a nice showing. Figure 5-96 shows an example of this balloon mail.

On June 4, MALÉRT made a Budapest-Gyöngyös-Budapest "first airmail flight," with cachets for both directions. This apparently had something to do with the *MOVE* glider club in Gyöngyös, and some of the cards bear a vignette of Arpád Molnár's advanced glider *Gyöngyös 33*. According to Short's book, no glider actually participated in carrying this

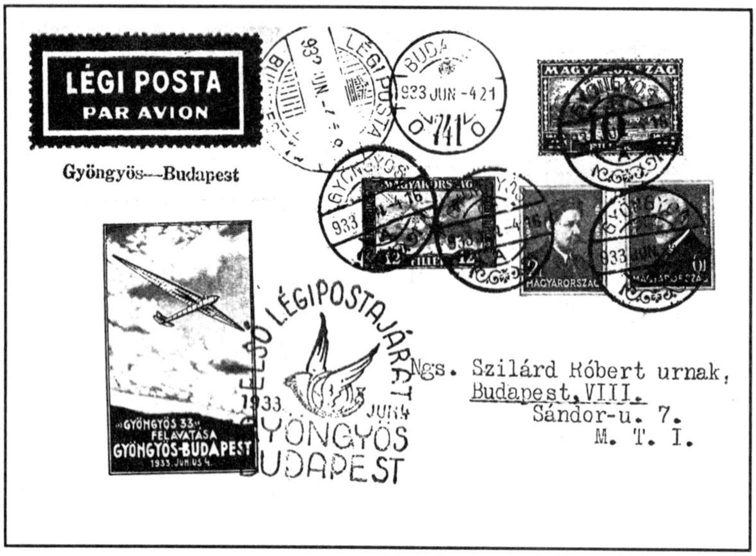

Figure 5-97 Gyöngyös-Budapest First Flight Card

airmail. Figure 5-97 shows a first flight card in the Gyöngyös-Budapest direction, with the glider vignette mentioned.

Finally, at the September 30 *LEHE 3rd Flying Day* event at Mátyásföld, cacheted postcards were carried aloft by *MALÉRT*.

1934 has five domestic airmail event commemorations. On May 9, in conjunction with LEHE's *2nd National Stamp Exhibit* and the *Budapest International Fair*, a one-day "experimental airmail route" was flown from Budapest to Debrecen and return. Violet handstamped cachets were applied. This airmail is notable for the frequent use of the beautiful Liszt souvenir sheet to frank it. Figure 5-98 shows a Debrecen-Budapest example.

June 24-5 featured a Budapest-Zalaegerszeg and return flight. The cachet showed the Boy Scout fleur-de-lis and a plane, and exists in both lilac and green, but only with the Zalaegerszeg-Budapest inscription. According to Short, Lajos Rotter's glider *Karakan* was towed by the plane, with the mail from Budapest being carried by the airplane and the mail from Zalaegerszeg carried by the glider. Some mail bears two frankings and was carried in both directions.

July 14, 1934 brings us to a Budapest-Miskolc-Nyiregyháza-Debrecen airmail flight sponsored by a Debrecen philatelic organ-ization. A variety of cachets and vignettes were used.

On August 5-12, the *Miskolc Stamp Club* sponsored airmail again for the Budapest-Miskolc-Nyiregyháza-Debrecen route. I consider this airmail, with its oversized color-of-the-day cachets, roller cancels, and gaudy vignettes, to be the ultimate in aerophilatelic bad taste. I mention it for completeness; but, have chosen *not* to picture any examples.

The final 1934 airmails to be mentioned are the simple, presentable cards flown at the September 27 *4th Flying Day* at Mátyásföld.

The only domestic airmail event of 1935 was the *5th FILPROK* exhibit balloon flights of June 4. [Note, the acronym for the exhibit was changed.] These "flights" were in "commemoration" of the 20th anniversary of the

Figure 5-98 Liszt Sheet on Debrecen-Budapest FFC

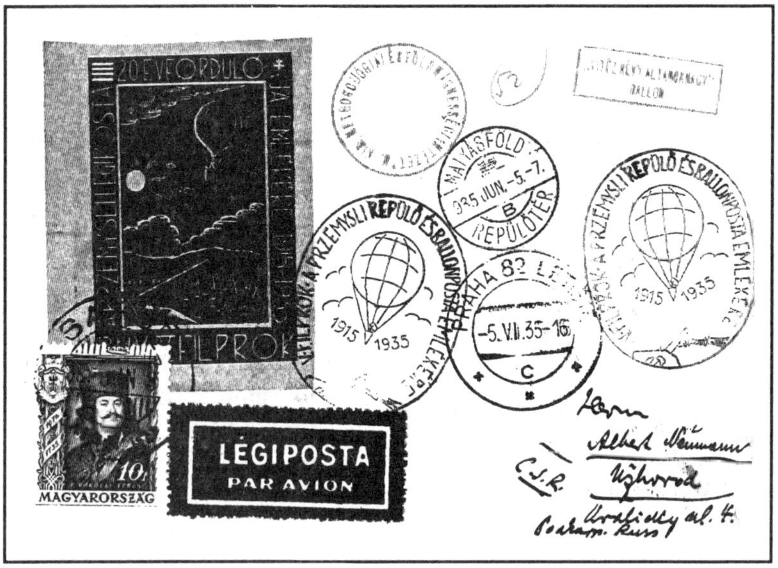

Figure 5–99 Przemysl Anniversary Balloon Mail of 1935

Przemysl balloon post. The four balloons were named: *Kuzmanek*, *Tamássy*, *Gyóni Géza*, and *Vitéz Révy*. True to form there were multiple cachets and vignettes. Figure 5–99 shows such a card flown on the *Vitéz Révy* and later flown by normal airmail to Czechoslovakia.

1936 begins with the *6th FILPROK* exhibit airmail to Dunaharaszti of May 9. Then the *Miskolc Stamp Club* took their turn with an August 15 Miskolc–Budapest flight. The large lilac cachet and vignettes continue their exploration of bad taste. Figure 5–100 shows an example of this Miskolc–Budapest airmail.

On a more serious note are the *Balaton International Sport Week* flights of September 5-8. Lake Balaton was developing into a major resort area. An airport at Balaton–Kiliti, just south of Siofok, was open to international air traffic during the summer, and a seaplane facility was in operation on the lake at Siofok. The airmail special event flights went from Budapest to Siofok, with drop mail to Balatonfüred enroute, and return to Budapest. This mail bears cachets, vignettes, and special receiving handstamps of Balatonfüred, Balaton–Kiliti, and (without a city name) Budapest. Figure 5–101 shows a 1936 *Balaton Sport Week* card dropped to Balatonfüred.

1937 domestic airmails started early with the *First Youth Balloon Mail* lightweight postcards of January. The balloon's name was *Solyom*, as indicated by a crude pink cachet. The imperforate vignettes are equally crude. The mail was posted on January 7 at Remete-kértváros. Since the stamp exhibit in Budapest that spawned this mail took place on January 1–3, it appears the cards were mailed after the unmanned balloon, which carried them, was found.

Figure 5–100 Miskolc Stamp Club 1936 Flight Card

FILPROK's 1937 contribution (their 7th exhibit) was the *1st Airmail Route* between Buda-pest and Szolnok. Large cachets featuring a posthorn and a variety of vignettes were featured.

An interesting aerophilatelic aside during 1937 is the July 5 postmark of *Budapest 20*. This was the first day of operation

Figure 5-101 Balaton Sport Week Airmail of 1936

of the post office at Budapest's new airport in Budaörs.

Latham also mentions June airmail of an *Air Rescue Conference* and a *Military Air Display* at Budaörs. I have seen neither of these.

The *Balaton International Sport Week* flights of September 11–15, 1937 covered a lot more area than that of the previous year. The airmail route this time was Budapest–Balatonkenese on the 11th, and then the *Balaton Air Rally* from Siofok on the 12th to Sopron, Keszthely, Kaposvár, Pécs, and Kalocsa returning to Budapest on the 15th.

To take advantage of all the additional stops, the organizing committee progressed to metallic (and other) vignettes in a variety of colors, and a circular origin cachet for each city. Figure 5-102 shows a large envelope picturing the entire route for the *Air Rally*. It was flown on the initial Budapest–Balatonkenese flight. [Unfortunately, the metallic vignette did not reproduce well.] Six cards -- one with each cachet -- makes a nice showing, but some collectors again look for the numerous combination routes.

In 1938, the first of the Vienna Awards was made by the Axis powers, returning territories to Hungary. The reacquisition of territories continued until 1941. Since the airmail associated with the *Returned Territories* is

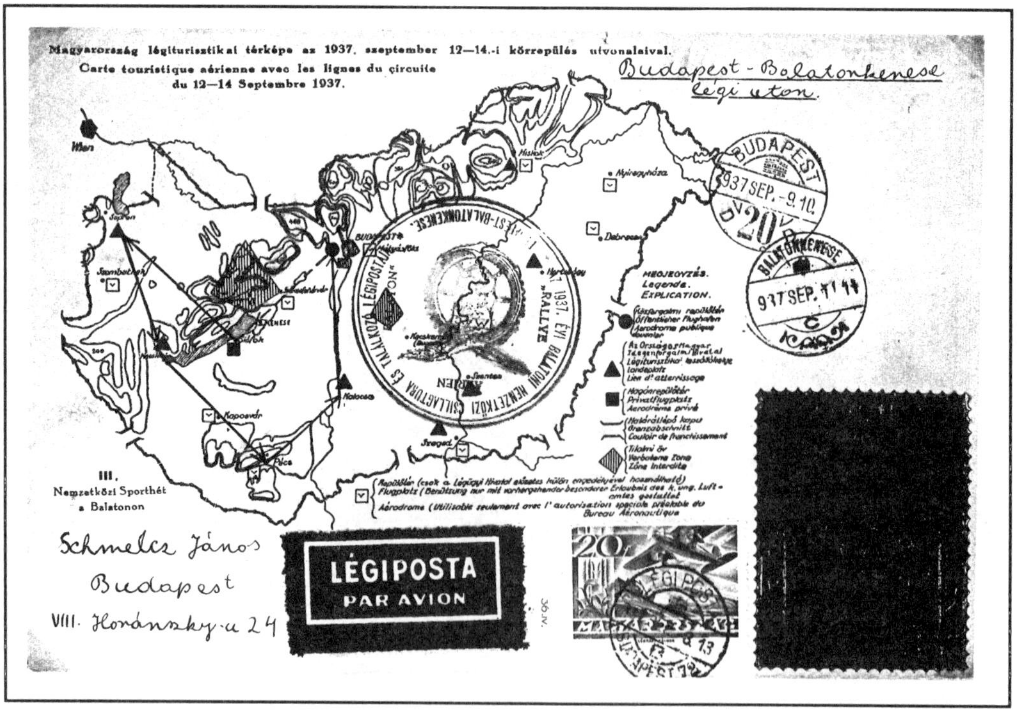

Figure 5-102 Balaton Sport Week Airmail of 1937

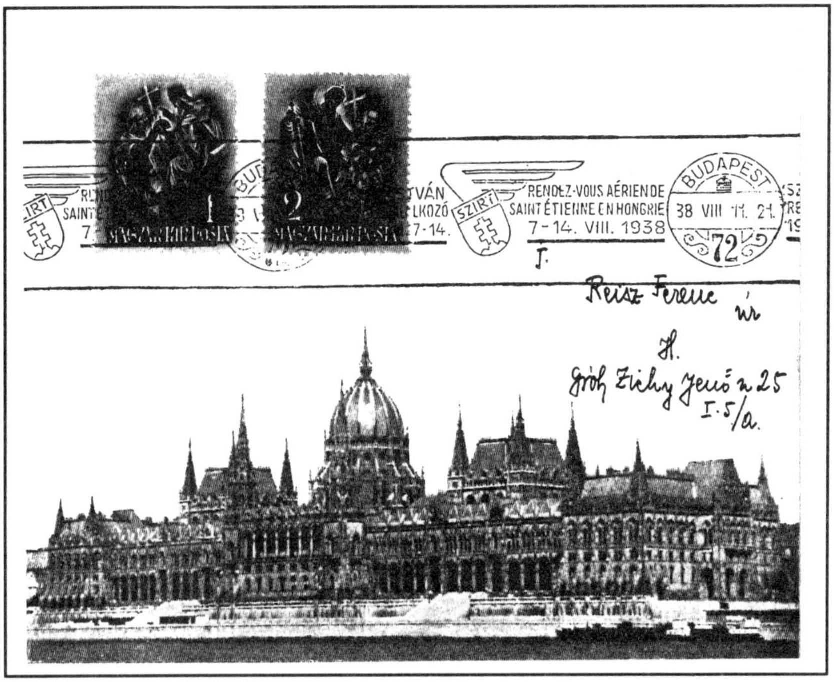

Figure 5–103 St. Stephen's Air Conference Cover

more appropriately covered in the context of World War II, this topic is deferred until the next section of this book.

1938 again presents us with a Budapest–Szolnok round–trip on July 3. No cachets this time, probably because it was a locally sponsored event. But, they did use triangular metallic vignettes, a special cancellation in Szolnok, and even customized airmail etiquettes.

The other airmail event of 1938 was the *8th FILPROK* Stamp Exhibit. This year, *FILPROK* sponsored four airmail flights to Czechoslovakia and Romania, but still had a place for a domestic Budapest to Esztergom and return flight. Cachets, vignettes, pictorial meters, and special cancellations were all included.

This year also provided two beautiful roller cancels relating to aviation and airmail, though they were usually used on normal surface mail. They are from the *St. Stephen Air Conference* held in Budapest from August 7–14, and a red "Postage Paid in Cash" roller cancel advertising for postal patrons to "Use the Airmails." Figure 5–103 shows the *Air Conference* pictorial cover.

The only significant domestic air event of 1939 is the *Scout Air Day* of May 21 at Szentes. This pictorial cancel, showing a glider, was used primarily on regular mail -- not airmail. Figure 5–104 shows the *Scout Flying Day* cancellation on an unflown cover to Colombia.

Since there is no other appropriate place, I will devote a few paragraphs here to the domestic "airmail events" of the early WW II period.

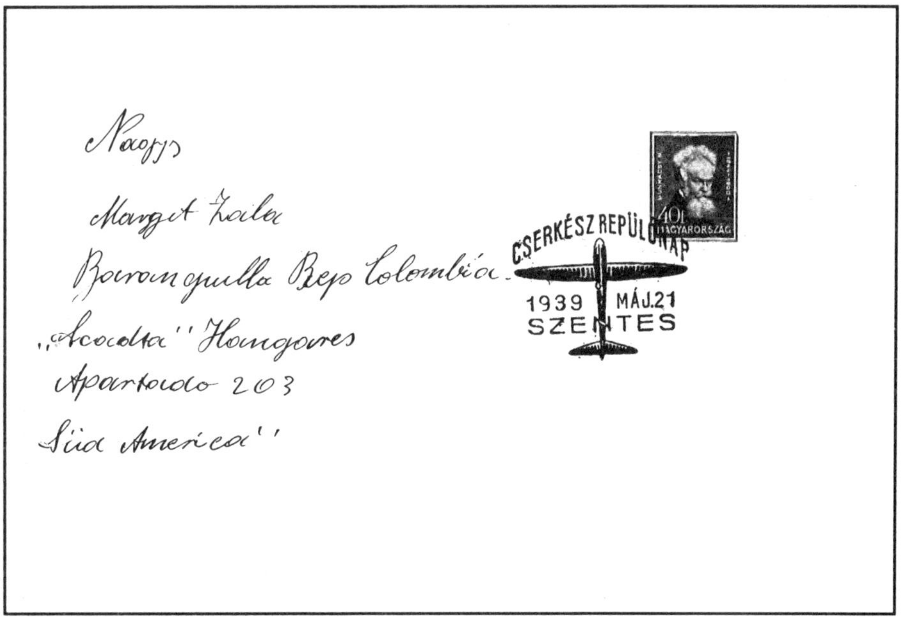

Figure 5–104 Boy Scout Flying Day Cancel

The last day of validity of the airmail issue of 1936 was March 31, 1939. With the coming of World War II, and the absence of airmail stamps, aerophilatelic events obviously took a "back seat," but a few were still to be found. These items often are franked with the *Horthy Aviation Fund* semipostal stamps with their flight motifs. It is very questionable whether any of this mail was actually flown.

Note: some airmail collectors also seek out the oval "Miklos Horthy National Aviation Fund" pictorial cancels. Actually, these are advertising cancellations for the semipostal stamps. So despite their aviation motif, these cancellations have no specific airmail significance.

Examples of airmail events in 1940 are the *9th FILPROK Exhibit* of April 7 commemorating the 25th anniversary of the Przemysl airmail service with large maximum cards, and the *MABOE* exhibit of May 19 commemorating the 100th anniversary of postage stamps with a special hexagonal airplane cancel.

FILPROK also celebrated their 10th exhibit on September 13, 1941 with a special cancel and a pictorial airmail meter. Latham mentions a few other examples in 1940 and 1941 that I have not seen. In December 1941, Hungary declared war and officially became a part of the Axis. That step ended all pretense of continuing domestic airmail events.

Airmail of the World War II Period

I have arbitrarily defined the "World War II Period" to begin on April 1, 1939 for a philatelic, rather than historic reason. March 31, 1939 was the last day of validity of the airmail issue of 1936. From that day, until 1947, Hungary had no airmail stamps. For that reason, the airmail of that era is neglected by the aerophilatelic community. It is an area that presents opportunities for research, for establishing a relatively inexpensive and interesting specialty collection, and even for augmenting that collection from dealers "junk boxes."

I like to start my exhibits of Hungarian World War II airmail with the unflown postcard shown in Figure 6–1. It is simply a 1936 airmail stamp canceled on its last day of validity. After this, you won't see another Hungarian airmail stamp in this book!

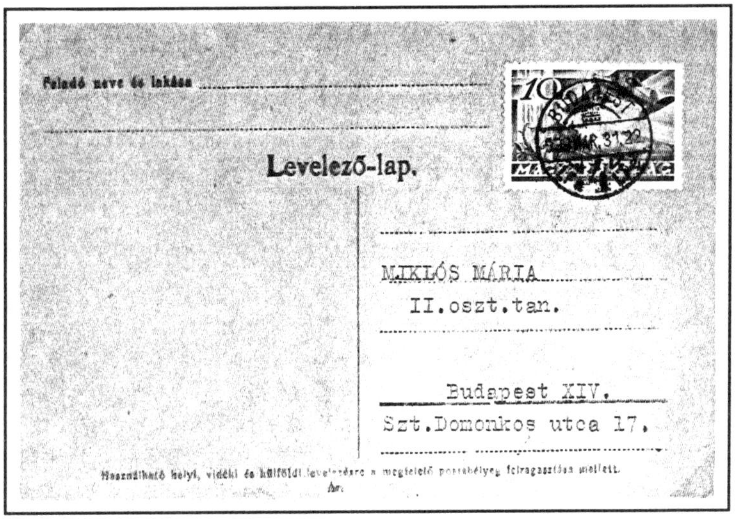

Figure 6–1 Last Day of Validity of 1936 Airmails

Airmail and the Returned Territories

The two–decade campaign to rejoin to Hungary some of the territories lost in the Treaty of Trianon was described in Section 1. This culminated in two "awards" and two annexations of land that are usually called by Hungarian philatelists "The Returned Territories." Figure 1–6 provides a map showing all the returned territories. In addition to the five major changes shown in the map, there were a couple of "minor border adjustments" between Hungary and Slovakia.

After each reacquisition of a returned territory, philately was used to make political "hay." Czechoslovak stamps, stamps of Slovakia, and Hungarian stamps were used separately or in mixed frankings on predominantly philatelic mail to commemorate the return to Hungary of many towns and cities. Sometimes normal cancellations were used, at other times special temporary cancellers were employed. Dozens of towns prepared special *Visszatért* ("returned") cancellations.

Little of this mail has any relevance to aerophilately; but, occasionally a few pieces are found that are of peripheral interest to airmail collectors. Figures 6–2 and 6–3 show two such examples.

Figure 6–2 Kassa Visszatért Cover

Figure 6–2 shows a November 1938 cover with the *Visszatért* cancellation of a major city -- Kassa. It bears a mixed franking of Hungarian regular and Czech airmail stamps, but is certainly unflown. [The U.S. stamp canceled by a New York World's Fair postmark was added two years later.]

Figure 6–3 is a 1939 airmail cover originating in Taksony -- a village that was returned to Hungary in 1939 as a part of a minor border readjustment with the new Slovak state. It is addressed to a Dutch stamp dealer who serviced covers from many returned towns. The card was flown only from Budapest to Rotterdam.

There were, though, two airmail routes established to the returned territories for short periods. The first -- to the northern areas of Slovakia and Carpathia -- is not documented and probably was only a single special flight.

Figure 6–4 Airmail to Northern Returned Territories

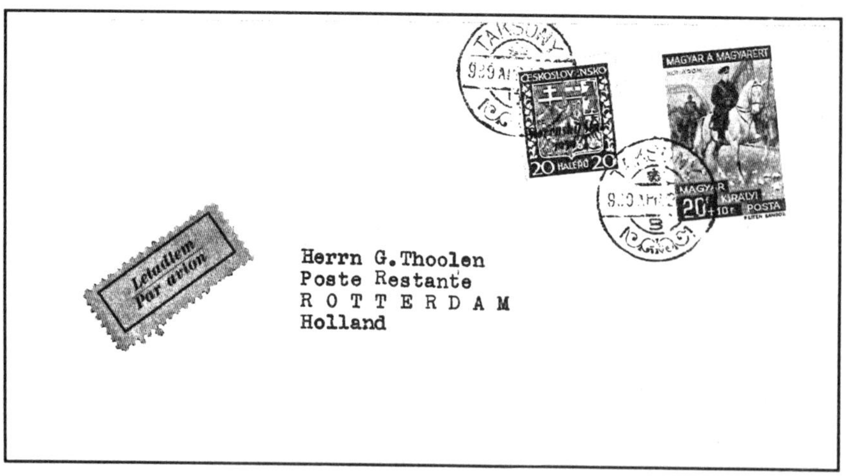

Figure 6–3 "Returned Territory" Airmail: Budapest–Rotterdam

On September 7, 1940, MALÉRT flew an airmail route from Budapest to Miskolc and then onward to Kassa and Ungvár in the northern territories. Figure 6–4 shows a card flown on the Budapest–Miskolc segment of this flight.

On September 24, 1940, the second route was established by

MALÉRT to the eastern territories, i.e. northern Transylvania. This route went from Budapest to Nagyvárad and Kolozsvár and was flown daily for several months, as reported by Jane's. Beginning October 7, an additional airmail destination was occasionally provided by parachute drops to Márosvásárhely in the Székely lands of eastern Transylvania.

This route was terminated with the crash of the Junkers JU–52 aircraft on landing in Nagyvárad on January 17, 1941. The burned and water–soaked mail from the crash was delivered to the addressees. One piece in the author's collection was posted in Budapest on January 16 and delivered in Szászrégen on January 22. It is shown, charred edges and all, in Figure 6–5.

for many years, although forbidden" under treaty. They also mention Hungary's "short private war with the Slovaks" in 1939, using primarily Italian aircraft.

Fieldpost of an "Air Regiment" is reported from the 1939 invasion of the Carpatho–Ukraine. While such mail was almost certainly not flown, it should be of at least as much interest to Hungarian aerophilatelists as the *Fliegerkompagnie* mail of World War I.

Jane's also reported that Hungary opened a new air academy in Kassa, in the returned territories, in late 1939. Little or none of this military aviation saga is recorded by Hungarian airmail. Probably, though, there is other mail "out there" that can shed light on this part of the aviation history of Hungary.

Figure 6–5 Mail from Crash at Nagyvárad in Transylvania

After this crash, there was no domestic airmail in Hungary until October of 1946.

One Hungarian aviation trend not reported by Jane's in the 1930's was the development of an Air Force, which was done largely under cover. Though a military air service became official in 1936, news of it to most of the world waited until the 1940 edition of Jane's where they note that "Hungary has had a small but well–trained Army Flying Corps

World War II International Airmail

This chapter is sub–divided into several parts. The first is the continuing story of *South Atlantic Airmail*. With the coming of the war, this route went into decline and by late 1941 was unavailable to Hungarian airmail. Except for the few Zeppelin trips noted in the previous section, *North Atlantic Airmail* only began in 1939. Mail flown between Hungary and the U.S. is quite common until the U.S.

Figure 6–6 August 1939 *Luft Hansa* Airmail to South America

entry into the war. Airmail from Hungary to Asia and Australia was always unusual. With the coming of the war and with most of the lines of communication broken, *Airmail to the East* became increasingly rare. *Airmail within Europe*, at least the part of Europe in the German sphere–of–influence, continued until the last months of the war. The subject of rates for Hungarian airmail during this period is very interesting. *Military Airmail* to and from Hungary is very rare, and is essentially an untouched research area.

South Atlantic Airmail. By 1939, the South American airmail pooling arrangement between Air France and Luft Hansa was well established. The over–water leg of the route from Africa to Brazil was being easily flown non–stop.

With the coming of the war, Luft Hansa immediately ceased operation of its South America route. Figure 6–6 shows a cover flown in August 1939 by the German airline on one of its last flights to South America ... Budapest, Berlin, Seville, Las Palmas, Bathurst, Recife, and Rio de Janiero in less than five days!

Air France continued to operate the route sporadically until mid–1940 when it also shut down its operations to South America.

In December 1939, the Italian airline *Linee Aeree Transcontinentali Italiane* (LATI) began operations to South America. For most of the next two years, LATI was the only practical carrier on the Africa–South

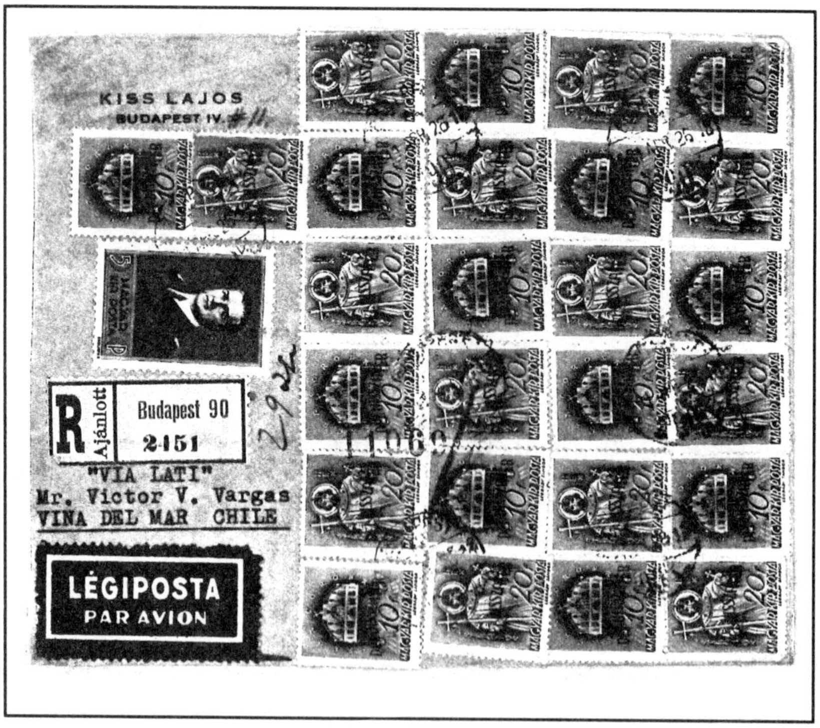

Figure 6–7 April 1941 *LATI* Airmail to South America

Figure 6–8 Omaha–Budapest Letter on First Trans–Atlantic Airmail

America route. Since regular air service continued between Hungary and Italy during this period, most, if not all, Hungarian airmail to South America during this period was flown by LATI.

Figure 6–7 shows an April 1941 airmail cover to Chile, flown by LATI. Hungarian airmail of 1940–41 to South America is scarce and getting quite expensive. Much of this mail bears very high frankings, indicating they were "money letters" -- probably moving cash out of the country prior to Hungary's direct involvement in the war.

In December 1941, the U.S. cut off all fuel supplies to LATI in South America and they ceased operation. Pam Am took over the wartime South America–Africa route as a contractor for the U.S. government.

North Atlantic Airmail.

Pan American Airways (Pan Am) began service across the North Atlantic, from New York to Marseille, via the Azores and Lisbon, on May 20, 1939. This service continued on a regular basis, but after France's entry into the war, Lisbon was the eastern terminus.

This "southern" route was supplemented by a "northern" route from New York to England, via Newfoundland and Ireland, in late June of 1939. Pan Am operated the northern route in a pooling arrangement with Imperial Airways.

Except for the first month, the route on which Hungarian airmail was carried is usually indistinguishable. The exception is mail that was censored by the British, which obviously implies the northern route.

The availability of these routes for Hungarian airmail ended when an official state of war was declared between Hungary and the Allies in December 1941.

Figure 6–8 shows a piece of American mail flown to Hungary on the first flight of the southern route. It was backstamped in Marseille on May 22 and then in Budapest on May 24.

Figure 6–9 is arguably the "last pre–war airmail" of Hungary. This badly damaged cover was in Paris, enroute from Budapest to New York, on August 30, 1939, the day before the war began. [The damage had nothing to do with the war. Just rough handling upon opening.]

I believe there is considerable research yet to be done on the airmail between Hungary and the U.S. in the May 1939 to December 1941 period. Issues such as the rationale for the use of the northern versus the southern route; censorship rules in Hungary, Britain, and elsewhere; and the detailed rate structure are all little understood. This represents a good opportunity for many collectors because prices are still reasonable; and so, the "purse" is not the controlling factor in such work.

Figure 6-9 Airmail Cover in Paris when the War Began.

To give an example, the cover in Figure 6-10 is difficult to understand. It is apparently a light-weight commercial cover postmarked July 1, 1941. It was censored by the British, so it traveled via the northern route. But, it has a previously unrecorded English-Hungarian *Transatlantic air-mail* handstamp and is franked 2 Pengó. Is it a private handstamp, or was it in wider use? How was the postage arrived at? It is very similar to other covers of the summer of 1941, with both lower and higher frankings! There's work to be done.

After the U.S. entry into the war, one possible route still existed for airmail to get from Hungary to the U.S. and other areas controlled by the Allies. That was *P.O. Box 506* in Lisbon, Portugal. The British company Thomas Cook operated this box as a neutral address for remailing letters between correspondents on opposite sides in the war. Its use for airmail from Hungary to the U.S. has been reported, but is extremely rare. In any case, it is unlikely that Thomas Cook forwarded mail received in Lisbon via air.

Figure 6-10 1941 Airmail to U.S.

Figure 6-11 is an interesting cover of June 1941. It is addressed to Ottawa, Canada and bears a Hungarian stamp canceled in Washington, D.C. Most people would have no idea what it is, no less why I proudly display it in an airmail collection. On the front is a handstamp reading *"This article originally mailed in country indicated by postage."* On the back is another reading *"Via pouch / H.K.T."* and a signature. What is it? It is a letter carried to the U.S. by a diplomatic courier and

Figure 6–11 To U.S. by Diplomatic Pouch

posted in Washington. It was probably carried as a favor for a Canadian diplomat in Budapest. Why airmail? Odds are the courier and his pouch flew aboard Pan Am's northern route that quickly became the route–of–choice for military and diplomatic dispatches, travel by British and American VIP's, etc. It is an unusual item, and peripheral to an airmail collection. But, it is the type of thing that will be found in an environment permeated by conflict.

Finally, adding to my picture of the sort of material one might look for in this period, Figure 6–12 shows an American airmail cover to Hungary. It was posted in New York on November 12, 1941. For reasons unknown, it was held up somewhere. It was censored in Hungary; that's perhaps the reason for the delay. On its reverse is a Budapest receiving mark of December 11. That's a few days after the war was on for both Hungary and the U.S. For nearly four years thereafter, no airmail, and very little other mail for that matter, traveled between the two countries.

Airmail to the East. Prior to the onset of the war, the most likely airmail carrier for mail from Hungary to Asia and Australia was KLM, not Imperial Airways as many people believe. Budapest was the eastern terminus of one Imperial Airways route. But, it was totally separate from the Imperial Airways route to Asia and Australia. So, mail carried by Imperial would have to go to London before being forwarded east.

On the other hand, the continuation of KLM's Amsterdam–Budapest–Athens route terminated (after many stopovers) in Batavia, Netherlands Indies (now Jakarta, Indonesia). From Batavia, *KNILM* (KLM's East Indies subsidiary), beginning in 1938, provided airmail service to Sydney.

Naturally, with the onset of war, circumstances changed. On September 2, 1939, all KLM services to the east from the Netherlands were suspended, including both routes to Hungary. For a

Figure 6–12 December 1941 Airmail to Hungary

short time, the European terminus of KLM's route to the Indies was Naples, Italy. In May of 1940, upon the invasion of the Netherlands by Germany, that terminus was again moved -- this time to Tel Aviv in Palestine. The KLM East Indies service continued to operate through Batavia, to Sydney, until February 1942. At that time, the Japanese capture of the Dutch East Indies caused the shutdown of all KLM operations in east Asia.

Figure 6-14 March 1941 Airmail to New Zealand

Notably, for about eight months in late 1939 and early 1940, Hungarian airmail could still go to parts of Asia and to Australia via KLM. This was because service from Hungary to Italy via Ala Littoria continued to operate regularly. Therefore, Hungarian airmail was carried to Naples by the Italian airline, and from there on to Batavia and Australia by KLM/KNILM.

Figure 6-13 shows a cover carried on that service. It was postmarked in Budapest on January 12, 1940, and bears a manuscript notation "via Italia." It is addressed to "Coogee, Sydney." Unfortunately for those interested in rates, this piece is no help. All the stamps were washed off when the KNILM aircraft crashed into the sea off Bali. The mail was salvaged, arrived at Sydney only one day late, and was handstamped on arrival in Australia to explain its condition.

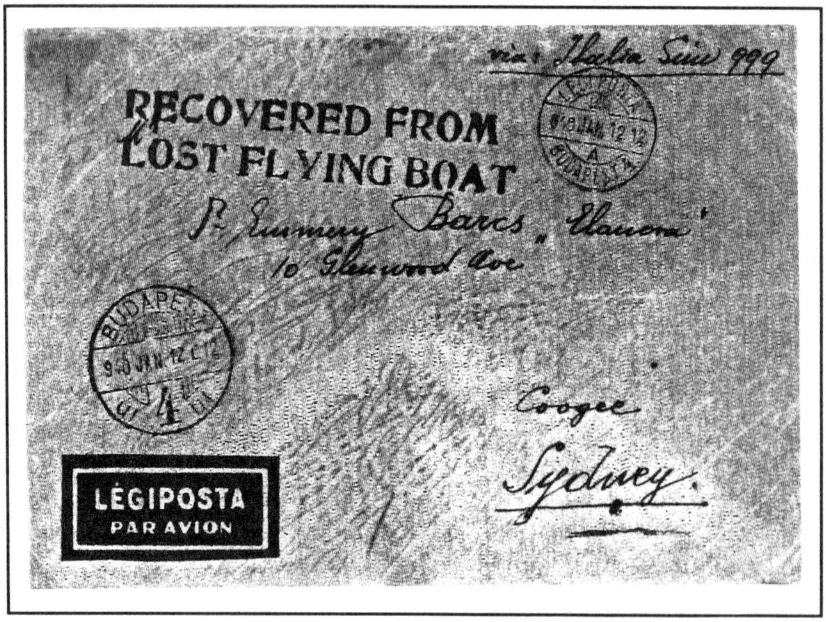

Figure 6-13 Crash Cover of 1940 Flown Hungary-Australia

What happened after May 1940? That's a good subject for research. Figure 6-14 shows a March 1941 cover from Hungary to New Zealand. Between the Budapest postmark and the New Zealand censor strip and handstamp, it has no postal markings that indicate its route. Most reported airmail of this sort is equally short of useful information.

Airmail Within Europe. As already noted, during the wartime period Hungarian airmail within Europe was mostly limited to the German sphere-of-influence and neutral nations (Ireland, Portugal, Spain, Sweden, and Switzerland). I have never seen wartime airmail to Ireland ... but the others all exist and are very interesting.

Figure 6-15 Airmail to Neutral Sweden

On September 15, 1941 Hungary raised its basic international postage rates for the first time in fifteen years. Three years later, German-occupied Hungary was being invaded by the Soviets and all airmail ceased. These three years provide many great opportunities to the study of a most unusual airmail rate structure to a variety of European nations.

The two key sources of information on postal rates during the wartime period are the "Monografia" and the "Handbook" -- see under Steiner and Surányi respectively in the bibliography. I will not attempt to reproduce that detail here. The essential idea is that there were two, and sometimes three, tiers of postage rates. It cost less to send mail to "preferred" countries. This was not only a question of proximity, but also of politics.

At one point mail to Germany had the same rates as domestic mail! Actually, the study of airmail rates between Germany and Hungary during the war would be a great research topic; and, it wouldn't be a very expensive project. Unfortunately though, most of the mail itself looks quite uninteresting.

On the other hand, mail to neutral nations and unusual destinations has a much more varied

Figure 6-16 Consular Airmail to Neutral Spain

and interesting appearance. But, other than Red Cross mail to Switzerland, it tends to be rather scarce and so very difficult to gather a comprehensive showing of rates and routes. I will illustrate four covers in this category here to give you an idea of the type of material that is available.

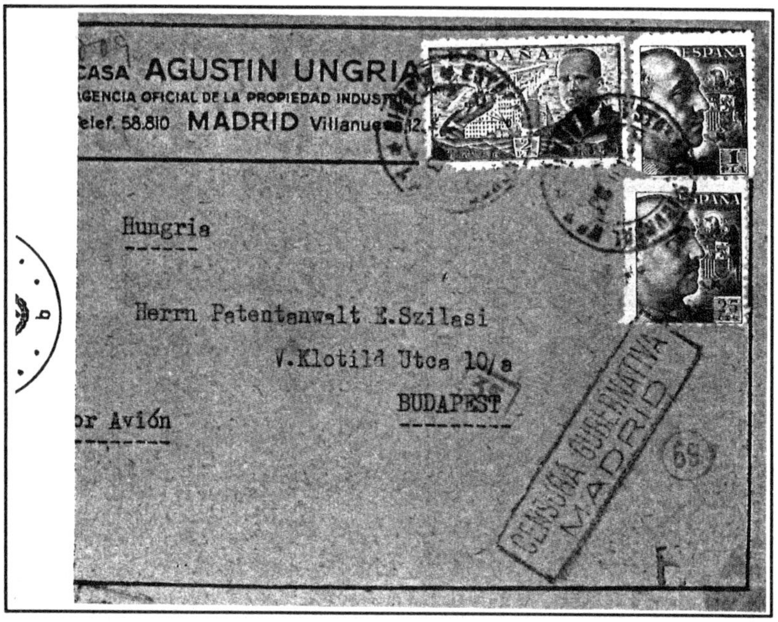

Figure 6–17 Spain–Hungary Airmail Censored Three Times

Figure 6–15 shows an April 1942 cover to Sweden. The rate is interesting. Postage and registry are 50 fillér each. The sender had to pay twice the 20 fillér airmail surcharge because the difference at that time between the airmail rates for "treaty" and "non–treaty" nations was the weight limit, not the amount of the surcharge.

Of course, virtually all international mail was censored, almost always by the Germans, usually by Hungarians, and sometimes by a third country. Figure 6–16 shows an April 1942 Hungarian airmail cover to the American embassy (the *enemy*!) in Madrid, Spain. It is unusual because it was *not* censored by either Hungary or Germany, probably due to the *KÖV* handstamp that was identified by Gazda as a marking for consular mail. It *was* censored by Spain.

Figure 6–17, on the other hand, is a June 1943 air letter from Madrid to Budapest that was censored three times, by Spain, Germany, and finally Hungary.

The final piece that I will show in this category is the February 1943 airmail letter from Hungary to occupied Norway you see in Figure 6–18. It shows the new special delivery rate that went into effect in September 1941. With multiple tiers, regular postage, airmail surcharges, registry, printed matter rates, special delivery, etc., showing a good selection of the wartime postal rates, especially on airmail, is a major challenge. I've found it to be worth the effort.

Figure 6–18 Airmail to Occupied Norway

Figure 6–19 Late Occupation Airmail from Hungary

Hungary's movement in the direction of the Axis is discussed in Section 1. From the time in 1941 when Hungary officially entered the war, she was a somewhat reluctant German ally. By 1943, Horthy's diplomats were discussing Hungary's unilateral withdrawal from the war with anyone who would listen. Hearing of this, Hitler summoned Horthy to Germany for "consultations." While he was absent, on March 19, 1944, the German army occupied its *ally* and installed a puppet government.

By this time, it was clear the war was going very badly for the Axis and airmail service from Budapest was less frequent. The Lufthansa corporate historian informed me that by the time of their October 4, 1944 quarterly report, all operations in Budapest had been terminated. Airmail of the occupation period is very scarce. I have never seen such airmail to any destination other than Germany, which of course included Austria at that time.

Figure 6–19 shows the latest airmail of the occupation period that I have seen. It is postmarked September 22. Keep searching; you may find one from a week later!

Military Airmail. If someone suggested collecting military airmail of World War II, you'd guess the "pickin's would be pretty sparse." You'd be correct, but there is more than you'd guess.

There is one type of official airmail *Tábori Posta* (fieldpost) stationery. It is an airmail reply card. Figure 6–20 shows a censored example from the front in the Soviet Union in November 1943. It is addressed to a military newspaper's office for "radio messages." It also has an additional "Legi Posta" handstamp. Whether it was actually flown, and if so where, is indeterminate.

As might be expected, I have found military airmail from a German soldier addressed to

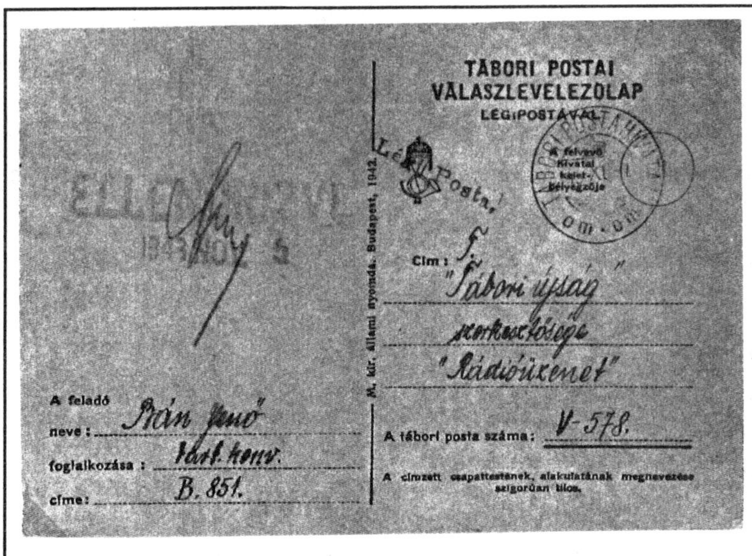

Figure 6–20 Soldier's Airmail Reply Card

someone in Hungary. Surprisingly though, I've also found that soldiers in Hungarian units attached to the German army could use German military airmail for messages home. Figure 6–21 shows an example that was censored by both the Hungarian and German armies.

Figure 6–21 Hungarian Usage of German Military Airmail

My final example of military airmail is a postcard from a German soldier in a military hospital in Kassa. Figure 6–22 shows the April 1944 card to Vienna. In addition to German censor marks, it has a German military airmail stamp canceled by a civilian Kassa postmark! I have no explanation of the usage, or whether it was flown. It does show though, that Germany was firmly in control of Hungary at that time.

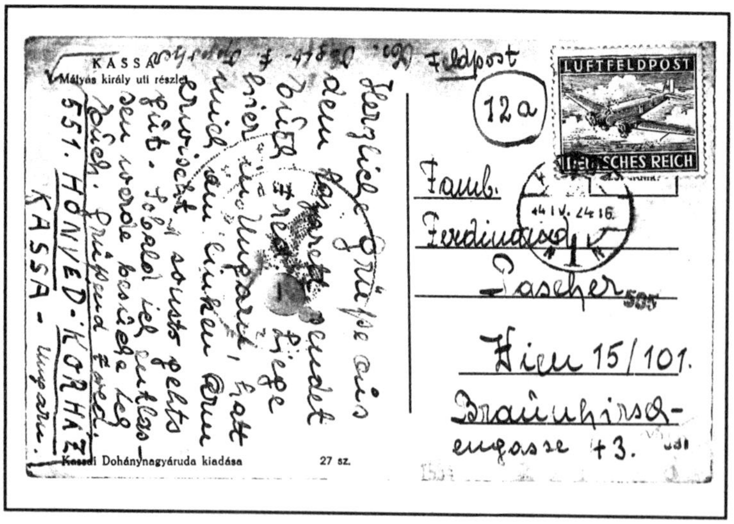

Figure 6–22 German Military Airmail Kassa–Vienna

1946 Airmail ... a New Beginning

From October 1944 to May 1946 no possibility for airmail existed in Hungary. An aviation enthusiast like me finds that unacceptable. To find a piece of mail related even to aviation was difficult. The postcard shown in Figure 6–23 has such a tenuous relationship to aviation I have hesitated to ever exhibit it.

The August 25, 1945 card is to the American Military Mission in Budapest. It is franked with 2 Pengó, the correct domestic postcard rate for the 2nd period of the inflation. The card is from a gravestone maker in Sárbogárd. He reports that he buried and provided gravestones for four American airmen killed in a crash near his village. We'll never know whether his was a humanitarian gesture, or if he was looking to be paid for his services.

The American fliers involved were from the 15th Air Force's *Operation FRANTIC*. Beginning June 2, 1944, B–17 *Flying Fortress* bombers, operating from Foggia, Italy, and escorted by P–51 *Mustangs* bombed Debrecen and other Hungarian cities. After releasing their munitions on Hungarian industrial and transportation targets, they flew on to airbases in the Ukraine. On the return flights, the *Operation FRANTIC* crews typically attacked Romanian oil fields.

At war's end, all aviation between Hungary and the outside world was banned, pending

Figure 6–23 Card Reporting American Grave Sites

conclusion of peace negotiations. Also, Hungary was experiencing the worst inflation in the history of the world. Between April 30, 1945 and July 31, 1946, the postage on a domestic letter increased by a factor of $8x10^{30}$ -- yes, that's an 8 followed by thirty zeroes! In that environment, it is a miracle that any airmail exists.

On February 4, 1946, KLM reinstated its service to Prague. On June 15, 1946, Pan Am extended its New York–London service to Vienna via Brussels and Prague. American requests that the Pan Am route be extended to Budapest were vetoed by the Soviet high command in Hungary. [This and other issues regarding post–war peace negotiations are discussed in detail in Stephen D. Kertesz' **Between Russia and the West**.]

Beginning May 15, 1946 airmail could be sent from Hungary via these routes. The mail was carried overland from Budapest to Prague, and flown from there by either KLM or Pan Am.

Users of this service had to pay the regular postage rate, plus an airmail surcharge that was payable only in gold francs. The surcharge within Europe was 10 centimes. For airmail to the United States the surcharge was one gold franc. The U.S. was the most common destination for such airmail, but even that is not at all common. Airmail to Australia or New Zealand paid a surcharge of two gold francs and is extremely rare. The surcharges were for mail of *only* five grams.

These airmail surcharges were enormous sums of money considering the circumstances in Hungary at the time. Obviously, air–mail letters carried via this service are quite scarce, and since the fees were paid in cash, also quite incon–spicuous. To the airmail specialist they are extremely valuable, as they are to inflation period collectors. Yet, being stampless and often in tattered condition, due to the very light weight paper used, they can on occasion still be found in dealers "junk boxes."

Probably all airmail to Europe and the East, and early airmail to the U.S. was carried by KLM. After the first month, Pan Am's service from Prague began and Hungarian airmail to the U.S. was probably carried by Pan Am, as they offered the most direct, and only through, service.

Figure 6–24 Prague–New York FF Cachet

I believe this theory is verified by a recently discovered June 14 Hungarian airmail cover addressed to the Bronx, NY. It bears the circular "Praha–New York" marking of June 17, shown in Figure 6–24. The cachet in this case was used as a transit mark. It was used only on Pan Am's first westbound flight from Prague on June 17. Capt. Albert Terwilleger piloted the flight Vienna to London via Prague and Brussels. The continuation to New York was flown by Capt. John Randall on the next day, June 18, via Ireland and Newfoundland.

A total of 1928 pieces of airmail were put aboard the plane at Prague. We'll never know the precise number of those that originated in Budapest, but covers of this flight are certainly among of the great rarities of Hungarian airmail.

Figure 6–25 15th Period Airmail to U.S.

1946 Hungarian airmail was almost always carried in very light weight envelopes. Remember, the weight limit was only five grams. Normally, these envelopes exhibit the following six postal markings:

1) *Légiposta / Par Avion* airmail etiquette;
2) black airmail postmark of Budapest 72;
3) red *Kézpénzzel Bérmentesitve* ("paid in cash") circular handstamp;
4) red rectangular *Taxe Perçue* ("postage paid" in French) marking, that is occasionally missing;
5) the weight of the letter indicated by a manuscript notation; and
6) the postage to be paid, also in manuscript.

A recent study by Szilagyi, published in the December 1995 issue of *News of Hungarian Philately*, found only 22 genuinely flown examples of these covers in all the major Hungarian inflation period collections outside of Hungary. [That study did not include the June 14 cover discussed above.] Of the covers in the study, three are addressed to England, one each to Australia and New Zealand. The rest are to the U.S. Since this mail all left Hungary in 1946, it is unlikely that significant numbers exist in collections there. These "ugly ducklings" are true rarities.

I show three examples. The first, Figure 6–25, was mailed to Los Angeles on May 16, during the 15th inflationary period. The weight indicated is 4½ grams. The postage was 540,000,000 Pengó.

Figure 6–26 shows a cover to Hamilton, New Zealand posted in Budapest on May 20, 1946. Its indicated charge is 2,160,000,000 Pengó and its weight four grams. The charge was a combination of the foreign letter rate of 160 million Pengó and twice the daily exchange rate of the gold franc, i.e. 1,000 million Pengó.

Because the "zeroes" had gotten totally out of hand, on July 12 Hungary switched to a new currency -- the *Adópengó* ("tax Pengó"). The rate of exchange was 1 AP = 2 quadrillion Pengó. Even so, postage rates rocketed. Figure 6–27 illustrates a five gram airmail cover mailed to New York on July 16th -- during the 25th inflationary period. It is correctly paid at 12,580,000 Adópengó or:

25,160,000,000,000,000,000,000 Pengó!

Figure 6–26 16th Period Airmail to New Zealand

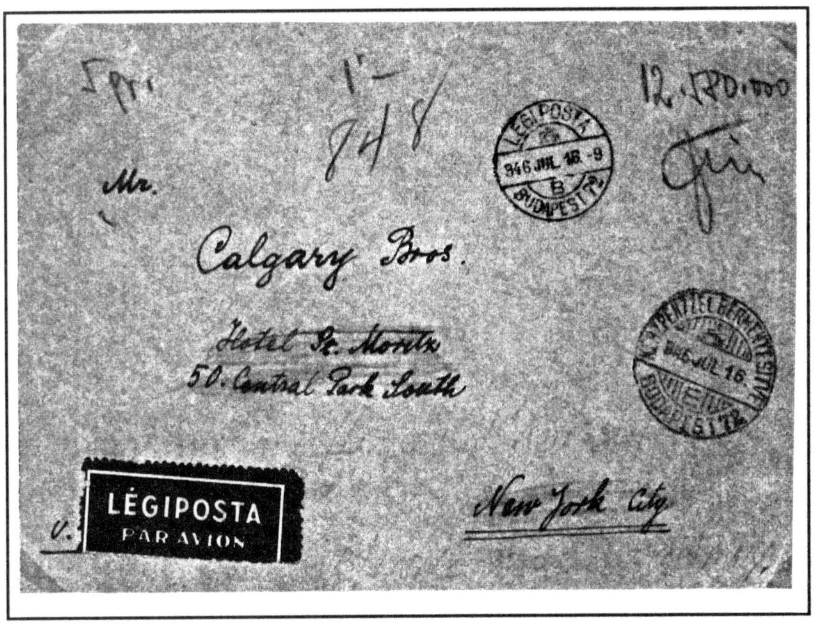

Figure 6–27 25th Period Airmail to U.S.

The temporary international airmail arrangements continued until the peace treaty. The Soviet Union was adamant that no air traffic could come to Hungary before. In August, the still-rare airmail looked very similar, but the payment was in the new currency, the Forint. Beginning in September, new stamps denominated in Forint came into use. When direct air service came again to Hungary in early 1947, airmail rates were reduced and Budapest was again a part of the world's airline network.

The first post–WWII domestic airmail was flown on October 15, 1946 when the joint Hungarian–Soviet airline *MASZOVLET* began internal air transport routes. These were the first of the long series of post-war political and quasi-philatelic events commemorated by airmail. But, domestic airmail in Hungary serving any real postal purpose was already long a thing of the past.

A peace treaty wasn't finalized until February 1947. An agreement by the Soviets to temper reparations demands and by the U.S. to free Hungarian assets, led to an end of Hungary's financial crisis. So, I end this story on July 31, 1946 -- the day the world's worst inflation ended in Hungary with currency reform.

Therefore, after 1946 Budapest was just another city in the routine handling of post-war airmail. The *Pioneer Period of Hungarian Airmail* had ended.

Annotated Bibliography

Note: The bibliography does not reference most of this author's previous publications on Hungarian airmail, since the book is intended to update and consolidate that information. For the curious, the author's previous works (in addition to the one monograph listed below) include about 40 articles in *News of Hungarian Philately*, *Stamps of Hungary*, *Hungarian Heritage Review*, the *Airpost Journal*, *Linn's Weekly Stamp News*, and the *American Philatelist*.

Alföldi, L. and Talas, S.; **Die ungarische Luftpost Handbuch und Katalog Band 1 1870–1930**; TAL AG.; St. Gallen SWI; 1993 (in German). This "volume 1" is all that has been published to date. The book is a compendium of material in the co-authors' outstanding collections combined with information taken from other sources ... primarily Latham. The coverage and photographs of the pre-WWI balloon mail are wonderful, and of the 1924–27 period also excellent. Many new and interesting discoveries are shown and theories put forward. There is surely more good work to come from these men. Available from the publisher.

American Air Mail Society; **American Air Mail Catalogue**; Fifth Edition;Volume 1; Washington USA; 1974 (in English). The section on Hungary in the category of *Early Foreign Flights* is poorly done and should not be relied upon for facts. The section on *Zeppelin Posts* is comprehensive and well-done. It is useful for anyone who prefers not to deal with the foreign-language Zeppelin catalogs. Likewise, the section on *Catapult Flights* in Volume 4 may be useful to English-only readers. Volume 4 also contains very useful information on FAM 18 ... which includes the 1946 route from Prague. Available from the publisher.

Berecz, Victor G.; **The Pioneer Period of Hungarian Airmail**; Society for Hungarian Philately; Fairfield USA; 1970 (in English). A small, rather dated monograph that evolved ... with a quarter century of additional research ... into this book with the same title. Out of print.

Clement, Alfred; **Militär-Luftpost 1793-1954**; Giorgio Migliavacca; Pavia ITA; Reprint Edition 1981 (in German); ISBN 88-85047-06-08. The classic source on the military airmails of the world. Good section on Przemysl, plus a few other interesting items for the Austro-Hungarian aerophilatelist. May still be available from the publisher.

Csanádi, Norbert, et al; **A Magyar Repülés Története**, Müszaki Könyvkiadó; Budapest HUN; 1977 (in Hungarian). An outstanding and detailed history of aviation in Hungary ... not perfect, but close. Particularly valuable for pictures and specifications of little-known aircraft built in Hungary. Availability unknown; but Davies told me in 1993 that an update is in the works.

Davies, R.E.G.: **A History of the World's Airlines**, 1983 (in English). Davies is now with the *Smithsonian Air and Space Museum* and is working on a follow-up book. This is an easy-to-read and outstanding volume covering the entire world, but with little specific to Hungary. Especially valuable are the sections on the beginnings of trans-Atlantic air routes, and on the *Imperial Airways* routes to Africa and Asia. It is out-of-print, but available in many public libraries.

Guidon, Louis; *Souvenirs de la Franco-Roumaine* in **Icare - revue de l'aviation francaise**, Paris FRA; Summer 1975 issue (in French). A detailed and well-illustrated history of CFRNA with a bit of a French perspective. A valuable reference, but by now may be difficult to find ... old magazine shops in Europe are probably the best bet.

Haberer, Erich; **Katalog über Die Katapultpost**; Rainer Haberer, Filderstadt GER; *Part 1 Nordatlantik*, 1985, *Part 2 Südatlantik*, 1988 (in German). The most detailed listings of Catapault Mail available. Invaluable for the specialist in that area, particularly for identifying specific flight dates. Available from the publisher.

Jane, Fred T. et al; **Jane's All the World's Aircraft,** published annually in the USA and UK (in English). This is the "bible" of aviation scholars because in each year's edition it records the key facts of military and civil aviation for every country in the world. I had access to the complete series from the first edition of 1909. I used the 1909–1950 editions in authoring this book. Recent editions are available in most libraries, but the complete run is only available in specialized aviation libraries. The very important 1909, 1913, and 1919 editions were reprinted about twenty years ago. These reprint editions can be found in many large public libraries.

Keszy–Harmath, Lorand; **Magyarország és a Zeppelin–járatok**; 1982 (in Hungarian). This is a nine–page magazine article which I acquired in the form of photocopy about 10 years ago. I am not sure where it was originally published, possibly *Philatelia*. It is an outstanding little catalog of Hungarian Zeppelin mail. Of particular interest is the author's approach to organizing the mail of the 1931 *Hungary Flight* of the Graf Zeppelin.

KLM Public Relations Bureau; **KLM Facts**; 1984 (in English). A nice little booklet giving a good overview of KLM's history, aircraft in use, and cities served ... including Budapest.

Kosel, Adolph; *Geschichte der Österreichischen Flugpost* in a special issue of **Austria–Philatelist**; Verlag Adolf Kosel; May 1948 (in German). Kosel was a prominent Vienna airmail dealer. Like Sobetzky, Kosel was the creator of many central European airmail "rarities." This little book is essentially a post–war update of Sobetzky's catalog, without the "Heil Hitler!" that was mandatory in 1938 Austria. Well organized, but not a lot of new information. Probably hard to find.

Latham, John J.; **Hungarian Airmail and Flights 1896–1976**; Magyar Philatelic Society of Great Britain; Oxford UK; 1977 (in English); ISSN 0307–4323. The best listing in English of Hungarian airmail covers. Its disadvantage is that it lists many normal airmail covers as if they were special flights. Two sets of addenda were published by MPS of GB in 1978–9. Possibly still available through the Hungarian Philatelic Society of Great Britain.

Latham, John J.; **Magyar Légiposta - Magyarországi Repülöjáratok 1896–1978**; MABÉOSZ; Budapest HUN; 1979 (in Hungarian). This is a translation of Latham's English–language book with the addenda and some other material incorporated. There are a few serious translation errors, probably due to Latham's inability to proofread the Hungarian version. Possibly still available through MABÉOSZ.

MALÉV Hungarian Airlines; **MALÉV Presents**; 1975 (in English). A history of the airline since 1945, with a short overview of pre–war aviation in Hungary. A few interesting facts, but tends to be politically motivated. Certainly out of print.

Rotem, Cvi; **David Schwarz -- Tragödie des Erfinders -- Zur Geschichte des Luftschiffes**; Privately published; Bloomington USA; 1983 (in German). This is the definitive history of Schwarz. It is a very limited edition and probably very hard to find. Try the University of Indiana library if you are interested.

Schupita, Peter; **Die k.u.k. Seeflieger -- Chronik und Dokumentation der österreichisch-ungarischen Marineluftwaffe 1911-1918**; Bernard & Graefe Verlag; Koblenz GER; 1983 (in German); ISBN 3-7637-5426-1. A remarkably detailed accounting ... down to every last aircraft ... of the history, mission, equipment, and bases of the Austro-Hungarian Seaplane Corps. It is available in specialist bookstores in the US.

Short, Simine; **Glider Mail -- An Aerophilatelic Handbook**; American Air Mail Society; Cinnaminson USA; 1987 (in English); ISBN 0-939429-00-4. This is the "bible" of glider mail. The section on Hungary is excellent, and the Austrian section also has information of value to Hungarian aerophilatelists.

Sieger, Hermann E.; **Zeppelin Post Katalog**; Sieger Verlag, Lorch GER; 21st Edition, 1995 (in German). I have not yet seen the 21st edition of this "bible" for the Zeppelin collector which was published very recently. My research was based on the 19th and 20th editions. The differences are likely to be simply the prices and the fact that a few more previously "unlisted" items are now listed. This is a "must-have" book for the Zeppelin collector. When I found my first Szt. András card, I turned to Sieger. The answers just "rolled out." Widely available.

Sobetzky, Georg; **Österreich-Flugpostkatalog**; Oesterr. Flugpostsammlerverein; Vienna AUS; 1938; 4th Edition (in German). This is the final edition of this catalog published by the greatest of all Austrian airmail collectors. It contains significant amounts of information of great interest to Hungarian collectors, particularly on the pre-WWI and wartime periods. It should be noted that in the 1930's Sobetzky was an Austrian postal official. He was not adverse to creating rarities for his collection, and he had the power to do so. This edition has a bit of a Nazi tinge, probably mandatory at that time in Austria. Xerox copies of this catalog have been widely available for years. Originals are probably very hard to come by.

Steiner, László et al; **A Magyar Bélyegek Monográfiája**; Volume 4; Közlekedési Dokumentációs Vállalat; Budapest HUN; 1971 (in Hungarian). Called by English-speakers the "Monografia," this is the "bible" of Hungarian philately. Volume 4 (of 6) contains most of the information of interest to aerophilatelists. There are some errors corrected a decade later by Surányi and Visnyovski's "Handbook." Only available on the resale market and getting expensive.

Surányi, László and Visnyovski, Gábor; **A Magyar Bélyegek Kézikönyve**; Gondolat; Budapest HUN; 1986 (in Hungarian); ISBN963-281-692-7. This "Handbook" is often referred to as the "black book" because of the color of its cover. It has an unbelievable amount of data in one small volume ... partially due to the thinness of the paper. It corrects a number of errors in the *Monográfiája*. Probably still available from Hungary.

Tombs, Laurence C.; **International Organization in European Air Transport**; Columbia University Press; New York; 1936 (in English). Extremely detailed data on the evolution of aviation in Europe, and on air transport treaties. To be found only in the most specialized aviation libraries.

Tranmer, Keith; **Austro-Hungarian A.P.O.'s. - 1914-1918**; Privately published; Hornchurch GBR; Revised Edition 1973 (in English). An excellent English-language source for information on the markings and locations of the various fieldpost offices during the war. Also includes good sections on Przemysl, the Naval Air Service on the Adriatic, and Air Force units. Was widely distributed after publication. Probably out of print.

Index

of references to personal names (both historical and philatelic).
Names of authors, when referencing or quoting their works, and personal names included in the designation of aircraft types and companies, are not indexed.

About the American Air Mail Society

GENERAL INFORMATION - The American Air Mail Society, organized in 1923, is one of the oldest and largest aerophilatelic societies in existence. AAMS members collect everything from the balloon posts of the siege of Paris to more modern airpost items, such as jet and rocket mail, contract air mail covers, transoceanic and foreign flight covers.

Dues are kept as low as possible and there are many privileges and services for members. It is not necessary that you be an advanced or wealthy specialist to attain membership, or to enjoy aerophilately to the fullest. The only requisites are that you have an impeccable character and have air mail collecting as an interest. Our members believe it to be an honor and a privilege to be affiliated with the group and often gather at stamp shows.

As a cooperative group organized solely to assist airmail collectors, the AAMS and all of its departments and services are not operated for profit.

THE AIR POST JOURNAL - This official monthly journal has been published by the AAMS since 1931. The *APJ* is a copiously illustrated magazine that is sent to each member for the duration of his or her membership. It covers the entire field of aerophilately and reports on new airmail stamp issues and new flights that occur throughout the world. AAMS membership news appears regularly as well.

CATALOGS AND HANDBOOKS - The AAMS has been the world's largest and most successful publisher of airmail literature. *The American Air Mail Catalog* has been published since 1935. Now in its fifth edition, a sixth is under preparation. Other handbooks treat a variety of specialist air mail topics. Now available are books on the siege of Paris in 1870-71, Newfoundland airmails, U.S. pioneer air mails and on many other subjects as well. AAMS members are entitled to a 20% discount off the usual price of these handbooks.

MEMBERSHIP SERVICES - Airmail stamps, covers and literature may be sold through the Sales Department or Auction Department. Exchanges with other AAMS members may be arranged through two (2) free members' want or exchange ads in *APJ*.

The AAMS provides for groups of collectors to form AAMS chapters that study special airmail topics. Some of these groups issue a newsletter as well.

A free translation service of many modern languages is available to all members.

The AAMS sponsors many awards for exhibits of airmail stamps and covers at regional, national and international stamp exhibitions. Certificates, medals and special awards are presented to airmail exhibits as decided by the show's judges.

ANNUAL CONVENTIONS AND MEETINGS - Two national meetings a year make it possible for members and friends to meet face-to-face and discuss their collecting interests. These meetings vividly display the simplest axiom of aerophilately- airmail stamps and covers remain the vehicles of our collecting interests, but the friends we make are the hobby's richest reward.

For information about any aspect of the American Air Mail Society write to:

THE AMERICAN AIR MAIL SOCIETY
P.O. BOX 110
MINEOLA, NY 11501-0110 U.S.A.